INTEGRATED SOFTWARE REUSE:
MANAGEMENT AND TECHNIQUES

Integrated Software Reuse: Management and Techniques

UNICOM

Applied Information Technology

edited by

Paul Walton and Neil Maiden

Routledge
Taylor & Francis Group

LONDON AND NEW YORK

First published 1993 by Ashgate Publishing

Reissued 2018 by Routledge
2 Park Square, Milton Park, Abingdon, Oxon, OX14 4RN
711 Third Avenue, New York, NY 10017, USA

Routledge is an imprint of the Taylor & Francis Group, an informa business

Copyright © UNICOM Seminars Ltd 1993

Notice:
Product or corporate names may be trademarks or registered trademarks, and are used only for identification and explanation without intent to infringe.

Publisher's Note
The publisher has gone to great lengths to ensure the quality of this reprint but points out that some imperfections in the original copies may be apparent.

Disclaimer
The publisher has made every effort to trace copyright holders and welcomes correspondence from those they have been unable to contact.

A Library of Congress record exists under LC control number: 93018658

ISBN 13: 978-1-138-31660-7 (hbk)
ISBN 13: 978-1-138-31666-9 (pbk)
ISBN 13: 978-0-429-45552-0 (ebk)

Contents

Preface

The collection of papers included in this book were given at a seminar organised jointly by UNICOM and the British Computer Society Software Reuse Specialist Group and held in London on the 3rd and 4th December 1991. The papers address a set of important, topical issues for the advancement of software reuse in the 1990s:

- effective management of software reuse, from motivating individual software developers to reuse rather than reinvent software components, to implementing organisational structures which promote organisation-wide reuse;
- recognition that many technical solutions to software reuse already exist;
- reusing system specification and design as well as code modules, thus supporting the critical but error-prone analysis and design phases of the software development life cycle;
- object-oriented approaches would appear to promote reuse, although little evidence of potential reuse is available;
- reuse of knowledge as well as software modules, typified by reuse paradigms developed in conjunction with domain analysis;
- critical success stories of software reuse in industrial-scale, real-world applications, encouraging the wider industrial uptake of software reuse;
- the need for metrics to measure potential benefits from software reuse;
- reverse engineering, the extraction of documentation and higher-level descriptions of software from the code itself;
- reuse of the software development process as well as artifacts from that process;
- the importance of human issues during software reuse tasks.

The papers in this collection vary from highly-focused academic papers, such as that by Martin Ward on formal specifications for reverse engineering, to the commercial, such as that by Kruzela. Taken together the papers in this book provide a wide spectrum of activity in current software engineering practice, especially that practice associated with promoting effective reuse.

The December 1991 seminar, from which these papers are taken, built on the success of two earlier seminars also organised by UNICOM in 1989 and 1990, whose papers are reported in the UNICOM Applied Information Technology Series book 'Software Reuse and Reverse Engineering in Practice', edited by Pat Hall (1991). The 1991 seminar proved to be the most successful to date, with higher attendance and greater participation from delegates, which is indicative of the increased recognition of software reuse as an important strategy for software engineering. The seminar was organised in collaboration with the British Computer Society Software Reuse Specialist Group, whose growth since its creation in 1989 is also indicative of the perceived future importance of software reuse by academics and industrialists alike. Increased home interest in software reuse has been matched by developments abroad. For several years, many large Japanese software houses have been implementing component-based software development, akin to their industrial manufacturing strategies. Similarly the existence of many software reuse workshops and tutorials in the United States has been symptomatic of the uptake of successful, large-scale software reuse programs by large US companies. This increased interest culminated in the First International Workshop on Software Reusability, held in Dortmund, Germany in July 1991, which will be followed by a second workshop in 1993. In short, software reuse has been recognised as a critical software engineering strategy for the future.

The papers in this book reflect the importance of integrating multi-faceted strategies for software reuse within software development organisations, hence the original seminar title 'Integrated Software Development with Reuse'. Previously reuse had been treated as a peripheral activity outside mainstream software development practice, so its uptake was slow. This book disagrees with this view and suggests instead that software reuse must be the cornerstone of effective software development practice. The introduction of object-oriented approaches, Computer-Aided Software Engineering (CASE) tool repositories and commercial code module libraries all indicate that reuse will be at the heart of future software engineering environments. However, many previous attempts to introduce software reuse into organisations have failed due to their focus on single reuse issues, such as retrieval or generification of modules or due to their mishandling of sensitive management and human issues, such as the Not Invented Here (NIH) syndrome. On the other hand, effective reuse can only be achieved through an integrated toolset assisting all software reuse tasks, and supported by a management which is committed to achieving software reuse in their organisation. In short, software reuse requires an integrated and comprehensive approach to be successful.

Such an approach must acknowledge the importance of two issues, effective management of the software reuse process and technical facilities which allow and promote software reuse. The apparent success of several large-scale reuse programs with modest technological support suggests that many technical problems inhibiting software reuse may already have been overcome. Instead, many of the problems appear to be managerial. Stories of previous reuse failures indicate that management must be fully committed to reuse for it to succeed. However, management commitment necessitates quantitative measure of potential benefits from reuse, especially in an era of cost-consciousness. As a starting point, this book presents several papers recording reuse success stories, and highlights inherent difficulties in demonstrating direct reuse benefits. Papers covering the technical aspects of software reuse look more to the future, and propose a battery of new and challenging techniques supporting both software and knowledge reuse. They suggest that the next generation of software reuse toolkits will represent a technological leap over current keyword-based approaches, founded on artificial intelligence, knowledge acquisition and human factors research integrated into software engineering environments, and offer even greater benefits from those offered by current software reuse research.

The papers in this volume cover a range of these issues and are divided into two broad categories: management papers and technical papers. The two categories are introduced and analysed by Walton and Sutcliffe respectively, with subsequent papers addressing specific areas in more detail.

Walton's paper outlines the main management issues associated with reuse. He begins with a section showing the benefits that have been gained form reuse programmes in a number of major organisations worldwide. In particular, significant benefits have been achieved in productivity, quality and timeliness. In each case the benefits have derived from a new approach to the whole software development process; one which aims to optimise the use made of organisational knowledge (by reusing it). The next section of the paper shows, in outline, the main roles and activities of this process, and the problems that can occur. The nature of the roles, activities and problems lead to the structural solutions described in Kruzela's paper. However, as well as organisational management issues, a new software development process generates new technical management issues which are discussed in the next section. The paper concludes with a summary of the discussion of the working group concerned with Managerial and Organisational issues at the First International Workshop on Software Reusability in Dortmund in July, 1991. The workshop was concerned with the pragmatic details associated with starting a reuse programme, and laid out a coherent approach, which is summarised in the paper.

Kruzela shows how the human problems associated with reuse can be solved. His paper contains a typical set of problems outlined by programmers and managers which indicate considerable resistance to reuse. These problems include the Not Invented Here (NIH) syndrome which has been much discussed. Kruzela uses examples of large scale reuse (from Toshiba, NTT and GTE) to show good organisational solutions which can be used to solve all of these problems.

Mole's paper examines the need to measure potential benefits in reusing software over developing it from scratch. Such measures are needed to support existing evidence and convince management and developers that reuse can work. Mole proposes several potentially effective measurements of software reuse drawn from existing software development metrics, although metrics specific to software reuse are also needed. Understanding how reuse can enhance software development can provide the motivation for its uptake on a large scale. One of the paths to software reuse which has been much discussed involves starting from existing systems and applying some form of reverse engineering to migrate to a different process. Reekie's paper discusses some of the difficulties involved in design recovery, and shows how some tools and techniques can be used to help the process.

Fowles discusses some of the key economic factors which underlie the management of software reuse component management. These factors operate within a market for components, where the market may operate within an organisation or externally. Fowles discusses some examples in which component management has been successful and analyses some of the issues involved.

Sutcliffe's paper provides a framework for investigating technical solutions to current problems in software reuse. He reviews the state of the art in software reuse by highlighting barriers to progress and technical solutions for overcoming these barriers. Current barriers to progress include repositories, failure to consider human factors during reuse, management of reuse and legal issues. The body of Sutcliffe's paper investigates four technical approaches to reuse, namely object orientation, formal methods, domain analysis and templates. Each is examined using a framework of technical issues covering the major phases of software reuse. Sutcliffe's conclusions are not promising, but he does identify several 'rays of hope' emerging from behind the software reuse cloud. In particular, he proposes knowledge-based support for templates which can help bridge the application gap and unlock the vast quantity of already developed applications for potential reuse.

Other papers examining technical issues in reuse can be examined using Sutcliffe's framework. Ward's paper investigates formal methods and transformations in software reuse. He proposes a theory for program transformation and refinement which has proved very powerful for the derivation of programs from specifications and the analysis of existing programs in software maintenance. This theory is combined with a metaprogramming language as a basis for constructing a library of reusable components.

Hall provides a review of domain analysis, including definitions of what domain analysis is, representations for domain knowledge, and methods and tools for achieving and supporting domain analysis. However, he concludes that domain analysis still presents a large number of unanswered questions. This is, Hall concludes, because domain analysis is of much wider interest than software reuse, and must encompass other issues, such as knowledge elicitation from artificial intelligence and generalisation of requirements analysis.

McParland proposes the reuse of application templates during software design, thus providing larger productivity gains than code-level reuse. One example of such an application template is a stock control system. These templates, McParland argues, can be sold by CASE or specialist vendors, or extracted from existing applications using reverse engineering techniques.

Maiden's paper on human issues in software reuse also investigates one of Sutcliffe's barriers to effective reuse. Maiden emphasises the need to introduce and support the software developer in the reuse process, especially when understanding and customising complex and unfamiliar chunks of software. This implies that we improve our understanding of how people reuse software to inform the design of tools which actively support software developers during reuse.

Finkelstein, Kramer and Hales report a case study used as a framework for a critical analysis of software process modelling, with important implications for how software processes can be reused to develop similar systems in the same way. The authors present implications for future research directions from their observations during this case study.

Ratcliffe's paper investigates the creation of reusable components using Ada. Ratcliffe highlights some of the Ada Reusability Guidelines produced from the Alvey funded Eclipse project needed to make Ada adequate for supporting software reuse.

Woodcock outlines a new trend in the CASE market - incorporating a reuse library in a CASE tool. He describes some of the features AD/Advantage, which is such a product.

1 An Introduction to Software Reuse Management

Paul Walton
MacDonald Dettwiler

1.1 INTRODUCTION

Sound management techniques and principles form the basis for successful reuse. Many authors have reported that the technical problems of current software reuse have been solved sufficiently to achieve significant benefits. Instead, the major challenge is to solve the managerial problems. The following quotes are contained in (Isoda, 1991), (Margono and Lindsay 1991), and (Tracz, 1988a):

> When we consider the technological obstacles [in software reuse], however, we find that most of them are caused by some managerial ones.
> . . .the major reuse barrier oftentimes has nothing to do with technology. Therefore, in order for a reuse effort to succeed, not only do we have to address technical issues, but we also need to tackle non-technical issues such as managerial and economic.
> Many good people have been led astray by assuming that the software reuse problem needs a technical solution. . . .if one looks at the most-often-stated reasons why software is not reused, the overwhelming majority of them may be classified as psychological, sociological, or economic.

Fortunately, there are a significant number of important examples of working software reuse in a number of major organisations around the world. These examples demonstrate that the managerial problems are also solved.

However, in these cases results have been achieved after an extensive review of the software development process, and a close look at the following underlying ideas:

- tackling organisational objectives using sound engineering principles
- defining the best process to achieve these objectives
- ensuring that the process is monitored, analysed and improved
- ensuring that the process makes the best use of scarce resources.

These ideas are expanded in this paper and in the subsequent papers.

There is a clear consensus developing about approaches to the management of reuse. In the First International Workshop of Software Reusability, the working group concerned with managerial and organisational issues (Basili, 1991a) achieved a common understanding, based on participation from Japan, the USA and Europe. This understanding is developed throughout the papers in this book. In particular, Kruzela shows the importance of structure, and in this paper we look at some of the underlying principles which govern the choice of solution. The following sections discuss:

- the stage of development of software reuse
- some principles underlying the management of reuse, and solutions
- some technical management issues
- how to develop a software reuse programme.

1.2 SOFTWARE REUSE TODAY

One of the best kept secrets of software reuse is the impressive range of major organisations who have reuse programmes. The following companies, at least, have reuse programmes:

IBM, NTT, AT&T, Toshiba, Motorola, Hewlett Packard, GTE, NEC, Fujitsu, Hitachi, BT, Mentor Graphics, PACTEL, US West, NobelTech, Intermetrics, Boeing, Ford Aerospace, General Dynamics, McDonnell Douglas, NASA, Hughes, Siemens, Kodak, GEC, CSC.

In addition to these a very large number of smaller companies have their own approaches. Some of the major organisations (and especially some of the Japanese) have been running a software reuse programme for more than a decade. The following summary of the benefits (from Cusumano, 1989 and Tracz, 1988) shows what they have achieved:

- Hitachi reduced the number of late projects from 72% to 7% in 4 years, and now averages 12% late
- Hitachi reduced the number of defects per machine in the field from 100 to 13 in 6 years
- Toshiba improved productivity (in terms of delivered lines of code) by a factor of nearly 3 in 9 years (including 48% reused code)
- at the same time in Toshiba, the number of defects per 1000 lines of code dropped from 7-20 to 2-3
- NEC improved productivity by 26% to 91%
- at the same time, NEC reduced defects by one third

- Fujitsu reduced the defect rate in operating system software by a factor of more than 10
- Fujitsu increased productivity, at the same time, by two thirds
- Raytheon Missile Systems Division increased productivity by up to 50%
- NobelTech (which used to be called Bofors) have more than doubled their productivity
- NobelTech expect future productivity improvements to be of a much higher level.

These results cannot be considered in isolation, and must be understood in the context of the particular conditions in each case. In particular, software reuse is not the only contributor, but it does form an an essential component. In each case, the key idea is that the software development process has been rethought. These organisations have independently concluded that the needs of the organisation are best served by pooling valuable pieces of knowledge (in the form of reusable components) and making that knowledge widely available.

1.3 THE MANAGEMENT OF REUSE

Software reuse is concerned with knowledge. Huge amounts of organisational knowledge are tied up in both the products and process of software development. Reuse is concerned with structuring this knowledge so that it makes the most effective contribution to the organisation's objectives. As expressed in (Barnes, 1991) 'Good reuse is...the reuse of human problem solving'. So potentially, any of the products of the process can be reused as well as parts of the process itself.

Managing these extra products adds new complexity to software development. The process must also cater for, and integrate:

- component lifecycles
- library lifecycles
- changed project lifecycles.

This complexity requires different management techniques and structures.

1.3.1 Roles

As a result of the changes in lifecycles, many new and changed roles occur in the reuse process. They are explored in (Basili, 1991), (Prieto-Díaz, 1991a), and (Kruzela, 1992). Broadly, and using the terminology in (Basili, 1991), the following roles are necessary to achieve reuse:

- definition of reuse policy and funding
- definition of materials/components to be reused (and how they evolve)
- domain analysis
- component design and development
- component maintenance
- component adaption and manipulation
- management of the library/experience base
- support, education and training for the reuse programme.

1.3.2 The reuse process

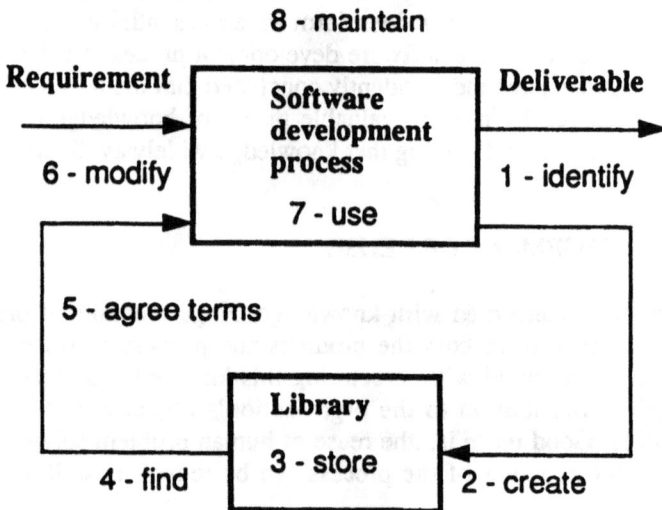

8 - maintain

Requirement

Software development process

Deliverable

6 - modify

7 - use

1 - identify

5 - agree terms

Library

3 - store

4 - find

2 - create

Figure 1.1 Reuse stages

These roles form part of the overall reuse process. Figure 1.1, which is adapted from (Kruzela, 1992), shows reuse in the context of the software development process. It shows the various stages of reuse.

Before anything can be reused, it has to be identified and created. After that it has to be stored in some form of library. The user must be able to search for, and find, the component (s)he wants. The buyer and owner must agree the terms associated with the use of the component. In many cases it may be necessary to modify the component before using it, and in the longer term the component will be maintained.

1.3.3 Reuse problems

Before trying to design the software development process, it is important to understand what the problems might be. These are summarised in Figure 1.2.

Reuse has a number of different potential problems, and the best way to understand them is to consider the moment at which the decision to reuse is made (or not).

A simple rule applies: reuse will occur if the benefits of reuse are greater than the difficulties/cost of reuse + the uncertainty. This implies to the person involved in reuse at the moment of reuse that the greater the certainty (of benefit), the greater the chance of reuse. Uncertainty is often time consuming and difficult to resolve, and can defeat any plans. The particular individual barriers are discussed in (Margono, 1991), (Kruzela, 1992), and (Frakes, 1991).

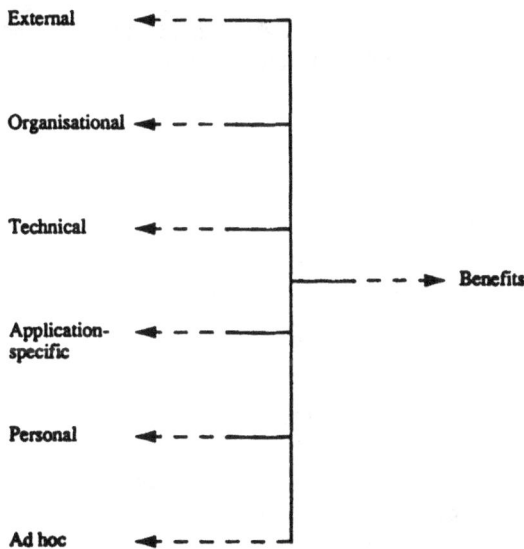

Figure 1.2 Reuse difficulties and benefits

The difficulties and costs fall into a number of categories.

External
If other organisations are involved in software development (for example if the software is produced under contract), then external difficulties may intervene. These are concerned with commercial and social factors. For example intellectual property rights and product liability laws may become relevant.

Organisational
If the supplier and user are in different parts of the organisation there may be organisational barriers to reuse. In general different parts of an organisation have different aims and objectives, and may not apply the same weight to an instance of reuse. In the worst case (which is very common), the supplier does not know that there is a suitable object to reuse somewhere else in the organisation.

Commercial difficulties can also surface. If the reuse systems are not a central overhead, or if they have to pay their way, internal pricing mechanisms may come into play. Warranty, support and maintenance will all have to be provided, and detailed terms and conditions agreed. One way to overcome these difficulties is to make the reuse system an integral part of software development as a central overhead, and bring all reuse projects under the same management. This may not fit with the current structure of an organisation, and may well involve considerable investment.

Technical

These problems are well understood. Any difference between the technical environments of the source and destination can kill any attempt at reuse. This difficulty breaks down into:
- machine differences
- development environment differences
- development method differences
- storage format differences.

These are usually overcome by making the source and destination environments the same.

Application-specific

Unless software is carefully constructed, it can contain implicit assumptions about the application it was developed for. This might make it hard to use it for different applications. This applies in general to larger pieces of code. The way to resolve this issue is through careful design, or to reuse software only within a particular domain or architecture.

Personal

The literature is full of personal difficulties concerned with reuse (for example, in (Kruzela, 1992)). From a managerial perspective, many of these are familiar problems. They can be solved through standard motivational techniques, but they may well have to be approached carefully. Education and training play a very large part.

Ad hoc

Finally there are many factors which are specific to each case. Some are purely administrative (Is it in the right (human) language? Is it in the right place?). Others are more fundamental (Was it designed for reuse?).

1.3.4 Measurement

The type of measurement used must relate to the particular objectives and requirements of the programme. Software metrics are of two kinds:

- general high level metrics that apply to all software development
- reuse specific metrics.

The first kind includes measurements of productivity, quality and timeliness, as well in some cases as adherence to standards. The second kind measures a number of aspects of the reuse process, including:

- how much is reused (by person, project or overall, and in absolute or relative terms)
- error profiles of old and new code
- adherence to the overall process
- reuse attempts (satisfied and unsatisfied)
- deposition levels (per person etc.)
- number of uses of components.

The Japanese, in particular, have produced highly instrumented processes with a wide variety of data collected. (Mole, 1992) describes metrics in more detail.

(Fowles, 1992) describes metrics and economic, productivity and quality models associated with reuse.

1.3.5 Solutions

[Kruzela, 92] describes a number of solutions using these management ideas.

1.4 TECHNICAL MANAGEMENT

1.4.1 Support systems

Figure 1.3 shows the technical issues which surround reuse, and which need support.

The whole software development process will need to produce management information for use in monitoring and controlling the process . The other issues are more geared towards reuse specifically.

Domain analysis is concerned with capturing knowledge about a domain (e.g. a type of application) in a usable form. Whatever techniques are used for capturing and storing the knowledge, it is important that it can be deployed easily to help assess what new components should be produced. Design for reuse is discussed in the next section.

Configuration management is becoming well understood and a number of products are available. The approach to take depends on the size of individual components.

Finding information in the library may be straightforward (if there are few items), and it may be sufficient to have the list of components made easily available. In more complex cases, with more components, more sophisticated approaches will be required.

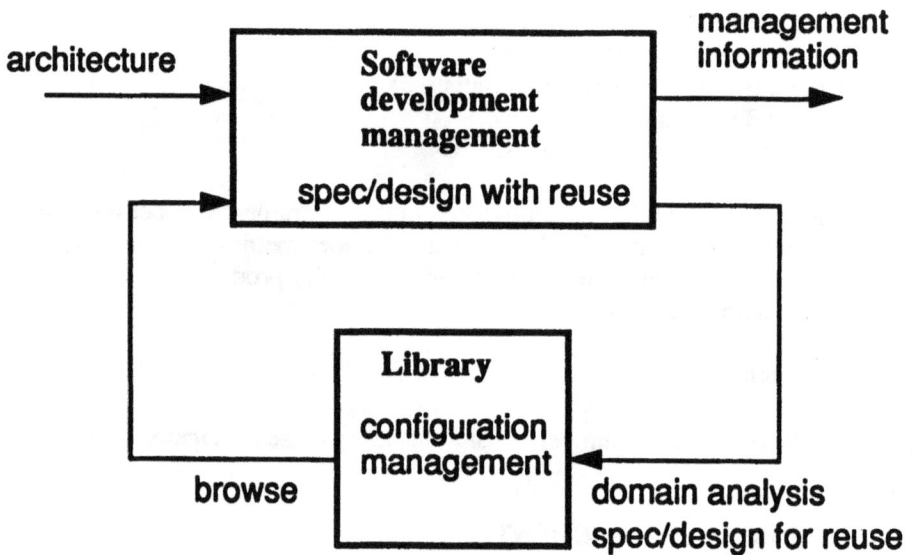

Figure 1.3 Technical considerations

1.4.2 Specification and design for reuse

'Reusable software', without further qualification, is a meaningless term. Software is reusable in some contexts and not in others, and so 'design for reuse' is not the important issue. Instead 'specification for reuse' is the key ingredient. The important element is a clear understanding of the range of alternatives which the component will address. The alternatives are of two kinds:

- application alternatives
  This represents the range of the application domain which the component must address
- platform alternatives
  This represents the range of hardware, operating system, CASE tool, software, which the component will run on.

In (Barnes, 1991) there is an outline of a more complex, probabilistic process, but the ideas are the same. Once these points are defined, the rest of the process (the 'design for reuse') is more straightforward. Reuse will stem from two fundamental design options:

- standard objectives into which different components will fit
- generic components.

After these choices are made, the rest of the component development uses traditional techniques. (Sutcliffe, 1992) describes these ideas in much greater detail.

1.5 INTRODUCING REUSE

This section is based on the discussion of the working group on Managerial and Organisational Issues at the First International Work shop on Software Reusability in Dortmund in July, 1991 (Prieto-Díaz, 1991).

1.5.1 Process improvement paradigm

The approach considered most profitable, and one which has been successfully used in practice, is based on a process improvement paradigm, using the following steps, applied cyclically. One of the main features of this approach is that this form of process definition and improvement embodies the best form of process reuse (regardless of any products created within it which can also be reused). The steps are:

- assessment
- defining objectives
- selecting a process
- planning the changes
- implementing changes
- measuring and analysing.

Once started, this process cycles continuously. Each of the elements of this process are described in the other papers, and many examples of software reuse programmes can be found in the literature.

1.5.2 Getting started

There are particular problems concerned with getting started. Once an initiative is rolling it acquires a momentum of its own. The initial assessment phase is extremely important. It should cover:

- the objectives of the software development process
- an analysis of the important software development assets of the organisation (of all types)
- current software developments techniques
- an analysis of reward structures and the motivation of individuals
- opportunities for pilot projects
- carrying out a cost/benefit analysis
- assessing the technology available.

This assessment should produce a clear picture of the opportunities and difficulties associated with the reuse programme.

To take the initiative any further clear management commitment is required at all relevant levels of management. The cost/benefit analysis will form an integral part of this, but it will also be necessary to indicate the levels of risk associated with the change. The fact that so many major organisations have an initiative can help here.

All the technical staff involved in the change also need to be committed. The approach to this is very organisation-specific, and depends on the culture. A critical element is the nature of reward and motivation in the organisation. Many issues are involved, including:

- career options available
- available incentives
- providing feedback.

Apart from these human issues, before the scheme is launched, a number of decisions have to be made concerning the nature of the reuse programme itself. Fortunately there are a number of good examples to base the programme on, but each organisation will need its own approach. The fundamental questions are:
- what material will be reused?

- where will it come from?
- who will create it?
- who will use it?
- how will people know what is available?
- how will the material be certified and supported?

The material for reuse can come from a variety of sources. It can be developed on projects, developed by a special division or it can be bought in. It can be based in existing software (which is re-engineered, re-packaged or documented), or developed from scratch. It will have to be stored in a library (or several libraries) and some systems set up to access the library. In order to determine what should be included in the library, some domain analysis may be required.

(Stehle, 1992) and (Reekie, 1992) show how reverse engineering can be carried out in practice. The library and systems will need the support of several organisational elements (as outlined in (Basili, 1991)) to carry out several roles, including:

- definition of reuse policy and funding
- definition of materials/components to be reused (and how they evolve)
- domain analysis
- component design and development
- component maintenance
- component adaption and manipulation
- management of the library/experience base
- support, education and training for the reuse programme.

These elements need to be integrated with other projects and software development, and this requires a thorough revision of the software development process. Software reuse ideas cannot be bolted on to the side of existing techniques.

1.5.3 Evolution

Once the programme is going, then the second and subsequent cycles of the process improvement paradigm take effect. According to the analysed metrics, a number of different options can be considered for the next steps of the programme. Some of the options are as follows:

- a different selection of reusable products
- a different number of reusable products
- application to a different domain
- application to more or different projects
- changes in the supporting structures and processes
- changes to the reuse process itself (perhaps to make it refer to different parts of the lifecycle)
- more or less formality in the process.

REFERENCES

Barnes, B. and Bollinger, T. Making Reuse Cost-Effective, IEEE Software, January 1991, pp. 13–24.

Basili, V. Presentation on Managerial and Organisational Issues to the First International Workshop on Software Reusability, Dortmund, July 1991.

Basili, V. Discussions of the Working Group on Managerial and Organisational Issues in the First International Workshop on Software Reusability, Dortmund,

July, 1991.

Cusumano, M. The Software Factory: A Historical Interpretation, IEEE Software, March 1989, pp. 23–30.

Frakes, B. A Survey of Software Reuse, Proceedings of the First International Workshop on Software Reusability, Dortmund, July 1991.

Isoda, S. An Experience of Software Reuse Activities, Proceedings of the First International Workshop on Software Reusability, Dortmund, July 1991.

Kruzela, I. Successful Management Structures for Reuse, this book.

Margono, J. and Lindsey, L. Software Reuse in the Air Traffic Control Advanced Automation System, Joint Symposia and Workshops: Improving the Software Process and Competitive Position, Alexandria, Virginia, USA, April-May 1991.

Mole, D. Measuring Reuse in Software Production, this book.

Prieto-Díaz, R. Proceedings of the First International Workshop on Software Reusability, Dortmund, July 1991.

Prieto-Díaz, R. Making Software Reuse Work; an Implementation Model, Proceedings of the First International Workshop on Software Reusability, Dortmund, July 1991.

Reekie, I. Migrating towards Software Reuse, this book.

Stehle, G. Reverse Engineering Studies, this book.

Tracz,W. Software Reuse: Emerging Technology, IEEE Tutorial, (ed.) W. Tracz 1988.

Tracz, W. Software Reuse Myths, in Software Reuse: Emerging Technology IEEE Tutorial, (ed.) W. Tracz 1988.

2 Successful Management Structures for Reuse

Ivan Kruzela
Telia Research AB

2.1 INTRODUCTION

Most people would agree that reuse in general is something 'good'. In the field of software engineering there are numerous papers praising advantages of reuse. Reuse can make production of software cheaper, faster, of higher quality, more predictable, more maintainable etc. The list of advantages can be made long.

Other papers demonstrate in a convincing way that reuse can be applied to all stages of the software life cycle. It is possible to reuse specifications, designs, test programs, all kinds of documents, the development process itself, knowledge, code etc. The potential for reuse of information of different kinds seems to be unlimited.

Good collections of papers on reuse supporting these claims are Freeman (1987), Tracz (1988), Biggerstaff (1989).

One of the great minds in the computer science field, P. Wegner (Wegner) 1983), in his enthusiasm for reuse goes as far as to say the following: 'Our desire to create reusable rather than transitory artifacts has aesthetic and intellectual as well as economic motivations and is part of man's desire for immortality. The drive to discover and exploit reusable patterns distinguishes man from other creatures and civilized from primitive societies.'

After such an overwhelming introduction, we must return to reality. If reuse is so great and so advantageous, why is it actually used so little in software production? If we honestly look around at the software developing companies we will see that reuse is done mostly in a spontaneous and often undisciplined way. Most of the time, reuse just happens, somebody simply remembers that he has earlier created something similar to what he is doing today and reuses it. Systematic reuse is still rare, except for reuse of large standardised components like window management systems or reuse in limited and well understood domains such as mathematical routines.

The reason for the not yet fulfilled promises of reuse is simple: planned and systematic reuse is difficult! The main advantage of reuse is that it reduces the complexity of the development process. But this reduction is not for free. Reuse has created many new complex problems that must be solved.

In the next section, we will list problem areas associated with reuse. In each area there is a need for more research and also more knowledge about practical experiences.

In Section 3 we will look more specifically into one of the areas. We will present a number of examples of human resistance to reuse. It seems odd, but this resistance is quite common and natural.

In Section 4 we will briefly describe three software developing organisations claiming to be successful. In Section 5 we will extract some common features of those organisations. Our contention is that the main factor to overcoming the natural human resistance to reuse is an organisation tailored for reuse.

2.2 PROBLEM AREAS IN REUSE

In a typical non-reuse based software production, requirements for certain deliverables, e.g. design, code or documents, are submitted to a development process which after a while, without regard for previous similar work, delivers a result.

In a reuse-oriented software production a useful and suitable piece of information developed in some previous activity is recognized and reused in the development process. It is as simple as that. The reuse leads to all the advantages which we mentioned in the introduction. But even though reuse simplifies the development process in many ways it creates new tasks and problems.

The simple model in Figure 2.1 provides a framework for identification of problem areas of reuse. Six general interdependent areas which include problems of a technical and non-technical nature are shown.

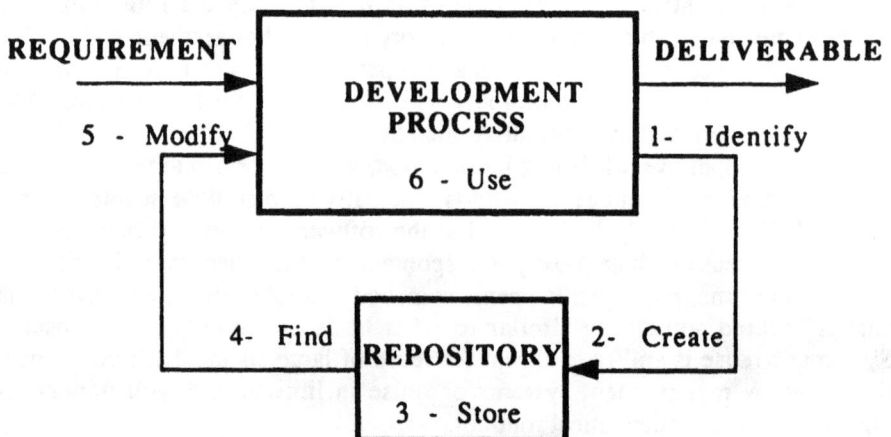

Figure 2.1 Problem areas in reuse

In principle, any kind of information occurring in the software development process at any place in a software producing organisation could be reused. But there are many trade-offs; it is not always true that everything that is reusable should also be reused. In the area (1) there are problems of *identifying* suitable reusable information.

The area (2) covers problems of how to *transform* reusable information into components. What is the proper size and form of reusable components and how should they be classified? What are proper techniques and languages for describing certain aspects of components? How to guarantee the quality of components? Formal methods will play an increasingly important role in the future.

The reusable components will be stored in a repository. Suitable types of *repositories* is the next important problem area (3). Is a relational database enough or is there a need for more powerful object-oriented databases with maybe some hypertext features? There are also some important non-technical problems in this area, e.g., the role of a librarian.

The future user of components does not have a complete knowledge about all the components in the repository and probably not even an exact understanding of what component he is looking for. So he needs *tools* for navigation in the repository and for retrieval and examination of selected subsets of components (4).

After obtaining a set of potentially useful components from the repository, the user must be able to understand, modify and use them (5). There is a need for *tools and methods* for evaluation and modification of components.

The area (6) is related to the development process which must support reuse. This area includes *non-technical aspects*, e.g. organisational, legal (copyrights, liability) and economical ones. The poorly understood domain of human aspects, discussed in the next section, can be referred to this area.

2.3 HUMAN RESISTANCE TO REUSE

Software production in large projects is a highly cooperative intellectual activity. Many persons with different roles are involved.

To simplify the discussion, we will consider only two roles in the software development process, a programmer and a manager. Both of them may be involved in two complementary activities related to reuse. The first activity is about *reusing* something and the other is *producing* something reusable. We will see that in both activities the programmer and the manager can have quite natural and legitimate reasons against reuse.

We will present attitudes of programmers and managers in the form of statements. Some of them were found in the literature but most were extracted from discussions with a large number of programmers and managers from different companies in different countries. There is no ranking between attitudes, some are more important than others. Some are probably more common in certain cultures and organisations.

2.3.1 Programmer's point of view

Using reusable components

- *I prefer to do it myself because it is more fun to write it than to just use another person's work.*
- *I will be showing weakness in front of my colleagues and my boss if I cannot do it by myself.*
- *It is not my code! I will do it better.*
- *It is easier for me to rewrite it than to modify it.*
- *It is difficult for me to maintain other people's work.*
- *I don't like the colleague who made this component and that is why I do not want to use it.*
- *I don't trust that person's components, he's always writing poor programs.*

Producing reusable components for others

- *I do not want to maintain my components used in a context I did not intend it for, which will be the case if they are used in another project.*
- *If somebody outside my project uses my component, he might discover its poor quality or the poor quality of its documentation.*
- *Software reuse is Software socialism! Why let anyone else benefit from my work and put my job security at risk.*

2.3.2 Manager's point of view

Using reusable components

- *The more lines of code that are produced the more impressive my project will be.*
- *A smaller staff can make my project more vulnerable.*
- *I do not have the full control of the quality of the components.*
- *The customer is my only friend, everyone else is a potential problem. I want to have full control of what is going on in my project and not waste time discussing with subcontractors.*
- *Reuse may make my project look too simple.*

Producing reusable components for others

- *If somebody else uses the results of my project, he might discover their poor quality.*
- *As a manager of my project I don't have any gains from long-term pay-offs of producing reusable components. Why bother wasting project resources with something that's not in the budget?*

2.4 REMEDY: ORGANISATION FOR REUSE

We present three organisations that are successfully practising reuse. They have
in common strong organisational commitment to reuse and an effective management structure. Reuse is promoted from the start of projects and is an organic
part of the software development life cycle.

2.4.1 Fuchu Software Factory

The Fuchu Software Factory (FSF), (Matsumoto 1987), manufactures application software for industrial process control systems in the Toshiba Fuchu works
of the Toshiba Corporation in Japan. FSF claims that the following impressive
figures of reuse rate are achieved: 32% of reuse in the design phase and 48% at
the coding level. High reuse rates are achieved also for other reusable items
such as test programs and documentation.

    The reason behind this successful reuse practice is mostly due to the organi-
sation depicted in Figure 2.2.

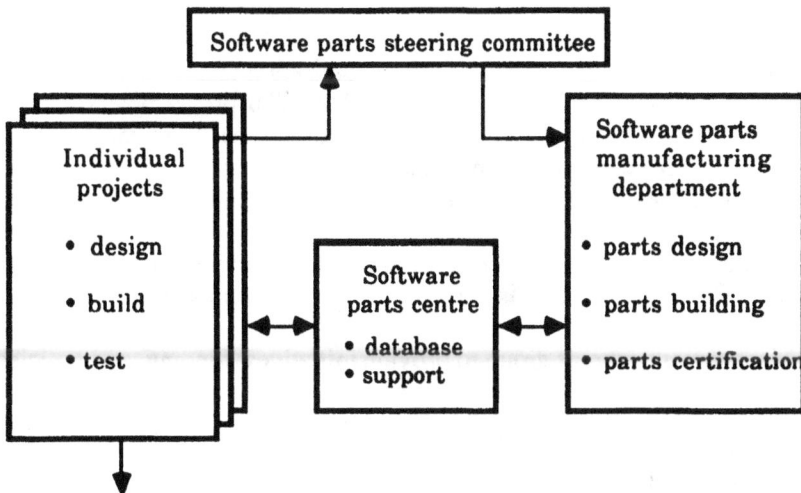

Figure 2.2

The three main components of the organisation are the following:

- software parts steering committee
- software parts manufacturing department
- software parts centre.

The steering committee gathers, selects and authorises the needs concerning creating, updating and discarding reusable parts. The reusable objects may be mod-

ularised documents, specifications and code. Different software tools, e.g., application generators are also reusable objects.

Parts are produced upon request by the manufacturing department. After thorough tests and evaluation the part is taken over by the software parts centre. To every authorised reusable part a check-list is attached which contains different evaluations of it. The software parts centre is responsible not only for the storage and registration but also provides services such as retrieval, maintenance and publication of reusable modules.

Software projects try to determine their need of modules as early as possible, preferably during the requirements capture phase. This early feedback from the software module centre provides a possibility for the project to influence the customer's requirement specification in order to take advantage of the use of existing parts.

A very important factor for promoting reuse at FSF is the regular training of programmers and designers in reuse and familiarisation with the content of the software parts centre.

2.4.2 GTE organisation

GTE Data Services in the USA have achieved code reuse rate of 14% in the first year (1987) after introducing reuse in the organisation (Prieto-Díaz 1990, 1991]. Considering software production volume (about 700 programmers) this translates into $1.5 million in savings. The ambitious goal is to reach 50% code reuse by the fifth year with estimated $10 million in savings. The increase of reuse will be achieved by the growing population of reusable components in the library.

An organisational infrastructure at GTE Data Services supporting reuse is depicted in Figure 2.3. It consists of the following components:

- Management support group to provide initiatives, funding, and policies for reuse.
- An accessible, densely populated, fully supported, and easy to use library system.
- An identification and qualification group responsible for the quality of the repository that identifies potential reusability areas and collects, procures and certifies new additions to the repository.
- A maintenance group that maintains and updates reusable components.
- A development group that creates new reusable components.
- A reuser support group that assists and trains reusers and runs tests and evaluations of reusable components.

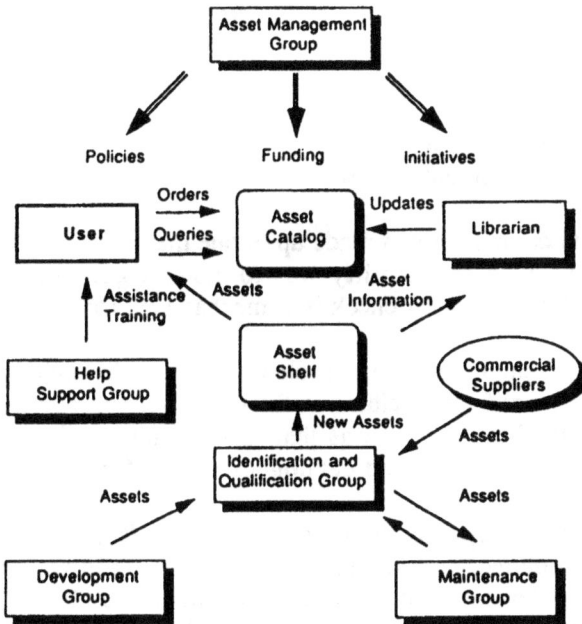

Figure 2.3 GTE reuse management

An interesting feature of the GTE method is that the company is giving very popular economical awards to inventors and developers of useful components. Programmers are paid $50 to $100 cash for each component accepted into the library and royalties are paid to program authors each time their components are used in new projects. A 'reuser of the month' award is given to those who reused the most. Managers whose projects achieved certain reuse level are given bonuses of budget extensions.

2.4.3 NTT Software Laboratories

The NTT (Nippon Telegraph and Telephone Corporation) Software Laboratories have performed an experimental reuse project lasting four years (Isoda, 1991). Its purpose was to gain experience in reuse, develop a reuse methodology and to determine the feasibility and limitations of reuse. Though the target of reuse was program code modules stored in a common library, the results and experience can be applied to other software artifacts including design, analysis and test cases.

The evaluation of the project is encouraging. The reuse ratio, i.e. the ratio of reused to newly developed software, of 15% was achieved. The average size of reused modules was about 600 LOC. After four years, the accumulated number of reusable modules in the library was one thousand. During the fourth year of the project 280 modules were used.

The project consisted of the following groups:

- reusability committee
- support group
- tool development group
- software development groups.

The Reusability committee was made up of one member of each software development group. Its responsibility was to discuss guidelines, standards for reusable modules, validation checklists, mechanisms for giving incentives and other matters.

The Tool development group developed some tools, such as a sophisticated module management facility, aiding in the search for appropriate modules. One surprising result of the project was that such an advanced tool is actually not needed, users preferred printed catalogues with information about available modules.

The Support group was guiding the project. It gave suggestions to the Reusability committee, reported to top management and was in charge of the construction and development of reusable modules and a library.

The Software development groups were the real workers of the project. Their software projects were chosen so that they had a good chance to reuse modules in the library.

Good component quality played a major role in creating confidence in using components. NTT developed a large number of requirements and recommendations to which software development groups had to adhere.

An interesting aspect of the project was a very popular award system. The most distinguished software development groups and individuals were given each year a number of prizes. Prizes included the high-deposition-ratio award, high-reuse-ratio award and highly-utilised-module award. The awards were not very large and they had more sociological than economic value. The reuse awards symbolize organisational effort, devotion and ideas of distinguished software engineers.

The project has been considered to be a success and the reuse technology acquired in the project has been transferred to other software development departments in NTT.

2.5 DISCUSSION

To make reuse a successful ingredient of the software development process it is important to consider both associated technical and non-technical problem areas. Problems having to do with human related aspects are often overlooked. In Section 3 we have listed a number of attitudes, some of which are perfectly natural, among actors in the software development field which make them less inclined to practise reuse. The sample of statements presented was about what is going on in people's minds. It is not important whether what people believe is true, false, right or wrong. That they believe it has consequences for their behaviour. It is a problem that must be dealt with.

We have briefly presented three successful reuse practising organisations. They have very much in common:

- Strong management support.
- Reuse is planned from the start of projects.
- Production of reusable components is separated from their use.
- Strong emphasis on good component quality.
- Repository of components with a dedicated support staff.
- Training of programmers in the use of the repository.
- An incentive to promote reuse in the organisation.

If we now study the negative statements in Section 3, we can see that most of them do not apply to Fuchu, GTE or NTT organisations. Users are not worried about the quality of components since components do have high and certified quality. Producers of components know that making components for others is their job and not an extra burden.

Our conclusion is that if a software development organisation wishes to introduce reuse it must start by creating the proper organisational environment.

ACKNOWLEDGEMENTS

This study was partly done within the ARISE project in the European Community research programme, RACE.

REFERENCES

Biggerstaff, T.J. and Perlis, A.J. (eds) Software Reusability, vol. I and II, ACM Press 1989

Freeman, P. ed. Tutorial: Software Reusability, IEEE Computer Society Press, 1987

Isoda, S. *An Experience of Software Reuse Activities,* Proceedings of the First International Workshop on Software Reusability, Dortmund, July 1991

Matsumoto, Y. *A Software Factory: An Overall Approach to Software Production*, Tutorial: Software Reusability, (ed.) Freeman, P. IEEE, 1987

Prieto-Diaz, R. *Implementing Faceted Classification for Software Reuse*, Proceedings of the 12th International Conference on Software Engineering, IEEE, 1990

Prieto-Diaz, R. *Implementing Faceted Classification for Software Reuse*, Communications of the ACM, **34**, (5), IEEE, 1991

Tracz, W. *Software Reuse: Motivators and Inhibitors*, Proceedings of COPCOM S´87, IEEE, 1987

Tracz, W. (ed.) Tutorial: Software Reuse: Emerging Technology, IEEE Computer Society Press, 1988

Wegner, P. *Varieties of reusability*, ITT Proceedings of the Workshop on Reusability in Programming, 1983

3 Measuring Reuse in Software Production

David Mole
CSSE, Southbank Polytechnic

3.1 INTRODUCTION

In competitive circumstances managers must strive to reduce the costs and improve the quality of the software produced. Reuse of software has long been advocated as a means of achieving these ends. In fact, software reuse has been practised since the early days of the Industry but this practice has not been so widespread as many would have wished. It is not appropriate to discuss all the reasons for this failure here but it appears that managers' lack of understanding of the potential for software reuse has been a contributory cause. In technology, what one cannot measure one does not understand. It is therefore worthwhile to examine some of the measurements that are relevant to the reuse of software.

We begin by examining the nature of software production. Then the uses of measurement in the management of software production are outlined. The difficulty of making useful measurements is illustrated.

Several ways of reusing software are described and the some of the measurements which can be made are discussed.

Several themes are stressed, viz. clarity of aims, consistency of method and minimising the cost of software measurement.

3.2 A VIEW OF SOFTWARE PRODUCTION

The software production process can be regarded as a system. From the viewpoint of the client (black-box view) the inputs are the client's requirements and the outputs are executable code and manuals. If the software already exists and is to be changed then the old software is also an input and the client expects to find much of it unchanged in the new software (e.g. commercial system maintenance).

From the viewpoint of the producer (glass-box) the production process is divided into a number of sub-processes each of which takes some document(s) as input (requirement, specification, design, test-data, etc.) and produces another document as output (specification, design, source-code, test-result, executable code, etc.). All these documents are regarded as software.

A general view of a such a sub-process is shown in Figure 3.1. There are various inputs and outputs, for example, persons, machines, consumables, overheads, source-code, object-code, manuals, test-sets.

Resources

Standards

Software (Documents) → **"Software Production Process" (A person writing !)** → **Software (Documents)**

Plans

Records

Figure 3.1 A view of the software production process

It is worthwhile to distinguish several kinds of input and output from the sub-process:

- software, for example designs and source-code;
- resources, for example persons, machines, overheads;
- control information, for example plans, standards and records of progress, including the measurements that we make on the other kinds of input and output.

A number of these sub-processes are joined together to make the software production process. The complete software production system may be characterised as a complex and labour-ntensive process for document production. The proper design and organisation of these sub-processes into a coherent whole is crucial to the success of the enterprise.

If the whole process is to be managed effectively then it is essential that it be well defined so that meaningful and repeatable measurements can be made. Making measurements on the process is the only objective way to set about improving it. All the documents, except the object-code, are written by persons: an inevitable consequence is that the 'process' of producing software (persons writing) is as variable as the persons involved. We cannot expect that measurements on software production will show the kind of repeatability that we get from a typical mechanical process like motor-car production. This variability is illustrated by the data in Appendix 1. Nevertheless, making measurements to facilitate planning, control and improvement of software production is well worthwhile. At least one company appears to have driven some of its competitors out of business by the management of software reuse for its products.

3.3 USE OF MEASUREMENT IN SOFTWARE PRODUCTION MANAGEMENT

Measurements are essential for planning, controlling and improving both the software production process and the quality of the product.

Planning requires estimates of both the resources required by, and the duration of, each stage of the software production. Unlike other engineering activities, there are no handbooks of data for estimating software projects, so that estimates for software projects can only be made from measurements on previous similar projects. Such estimates are notoriously inaccurate, one reason being the inadequacy of the historical data available to the estimator. Another reason, already noted, is the comparatively large variation between the performance of apparently similar individuals. Yet another difficulty of estimation lies in establishing a reliable correlation between the requirements for the software to be produced and the effort required to produce it. All these difficulties are much reduced for software which is reused. Even with reuse, measurement is essential for good planning and it is important that the collection of data for this purpose is carefully considered because one cannot usually go back to collect any items which were overlooked.

Measurements can be used to control production by comparison of actual and planned performance; this aspect of measurement is too well known to require more discussion here.

Provided that we define quality in measurable terms then measurements can be used to control software quality. From the client's viewpoint two important indicators of quality are conformance to requirements and the error rate in service. Both can be measured and the error rate can even be predicted from suitable measurements.

Measurement can also be used as an aid to understanding the process as the example in Appendix 1 shows.

3.4 MEASUREMENT

To measure successfully we need standards, tools for measuring and a little theory of measurement. For example, we have standard measures of duration and of cost, internationally agreed. We have very precise tools for measuring them (clocks, coins). We have theories for both (they are ratio measurements) which are so well understood that we scarcely notice that the theories exist.

We do not yet have agreed international, or even national, standard measures of software. For example size is often reported as so many lines of code (loc). These lines may, or may not, include comments and blank lines. The control structure and calling structure of source code may be measured but, again, there are no standards. So it is impossible to compare your software with others, except in terms of cost and duration of development. It is thus very important to set internal standards so that you can use your measurements for comparison with other earlier or later measurements. In this way you may hope to learn something useful.

The theory of measurement is worth some attention because the kind of measurement made determines the kinds of calculation that can be made with the data and therefore the kinds of analysis that can produced. For example, a mean value can only be calculated for interval, ratio, and absolute data. (For a good account of software measurement in theory and practice see N. E. Fenton, *Software Metrics, a Rigorous Approach*, Chapman and Hall, 1991.)

There are a large number of tools for software measurement on the market. A tool is worth having if it measures what you want to know. Having a tool does not of itself produce any useful results.

3.4.1 Software measurement

Above all it is essential to know why one wishes to make a measurement. Failure to clarify the objectives of a measurement programme will not only fail to produce results but will also waste resources. A famous statistician said, 'Data should be collected with a clear purpose in mind. Not only a clear purpose but a clear idea as to the precise way in which they will be analysed so as to yield the desired information. ... It is astonishing that men, who in other respects are clear sighted, will collect absolute hotch-potches of data in the blithe and uncritical belief that analysis can get something useful out of it.' (M. J. Moroney, *Facts from Figures* (third edition), p.120, Penguin Books, 1956) Software managers who, unlike many managers in other engineering disciplines, often have little training in quantitative numerate management, perhaps need to take this to heart.

If measurements on the software production process are to be of value then, before commencing to make the measurements, one must at least carry out the following steps:

- Produce a clear statement about the goal(s) for making the measurements.
- Identify clearly the objects to be measured (version control may be important here).
- Identify clearly what attributes are to be measured (cost, size, faults etc.).
- Decide what data to collect.
- Decide how the data will be analysed (by project, module, person, product etc.).
- Devise and install a data-collection scheme (coding, procedures, inform staff, recording).

3.5 MEASUREMENTS OF REUSE

3.5.1 Definition of software reuse

Care must be taken with the definition of reuse if we are to measure it successfully. Software reuse is the reuse of text and so we need to consider the various ways in which text is reused in software production. There are at least three ways in which we can reuse text:

- Copy unchanged, here there is only one difficulty – we may have to decide how we treat the case of multiple copies of the same text into one product document.
- Substitute and copy, for example by macro processor or a program generator that generates source-code.
- Amend (delete and insert text) an existing document and then use the amended version.

In all these cases there are no generally accepted standard measures so you must define your own. It would be sensible to use simple measurements like the fraction of the final product which is new (e.g. lines of manual, lines of code).

3.5.2 Measurement goals related to reuse

Any measurements that can be made on software produced without reuse can be made on software produced with reuse and also on libraries of modules for reuse. Here we consider only software measurements that are special to reuse.

The main purposes of software reuse are to decrease the cost of software production and to increase the quality of the product. Measurements of cost and productivity are straightforward. Measurement of quality is less common.

There are however some other measurements which are required for the management of software reuse. These are measurements made on software which incorporates reused items and measurements made on the 'library' – this term is used here for any collection of software for reuse, no matter how organised.

3.5.2.1 Productivity

For software which incorporates reused items it is important to measure both the cost of incorporating the reused items into the finished software and the increase in productivity. This information can be compared with the cost of writing the 'new' parts of the software and may also suggest ways in which reuse could be improved. The cost data can also be used for estimation of future products. The recording and analysis of this data can use standard techniques. Care should be taken to see that the size of the software is measured in the same way for both the 'new' and 'reused' parts.

3.5.2.2 Quality

The ISO Draft Standard identifies Functionality, Reliability, Usability, Efficiency, Maintainability, Portability as six sub-characteristics of Software Quality. Each of these may then be further subdivided into measurable characteristics, for example, Reliability may be subdivided into:

- fault frequency – can be measured as faults per 1000 hours or per 1000 transactions processed or as mtbf;
- fault tolerance – can be measured as crashes per 100 faults;
- fault density – can be measured as faults/1000 loc.

Faults should be recorded and attributed to either the 'new' or 'reused' parts of the software. Faults-in-service are one important aspect of quality. One expects an improvement in software quality as a result of reusing well-tried software and this should be confirmed by measurement. Any other attribute of quality which is important in your market should similarly be measured and attributed to the 'new' or 'reused' parts of the product.

3.5.2.3 Measurements on the reuse 'Library'

The measurements which are unique to software reuse are those that relate to the library of components for reuse. 'Library' is used here to mean any collection of components that are reused regardless of the way that they are organised. If one wishes to monitor and control the reuse of the items in a software library then for each item in the library we need

- the item to be reused (the text)
- the specification (description) of the item
- a record of the reuse of the item (say, date and project-reference)
- a record of changes to the item (say, date and type of change).

The specifications normally enable the users to choose the most suitable item from the library for their purpose. The specification must also be used to decide if there is a fault in the library item and so it is important that it is adequate for that purpose too.

The record of changes to the text should include the reason for the change (fault, design change, efficiency, etc.). The records of reuse and of changes will enable the manager to decide if the item should remain in the library (not used), or if it should be changed (much used, many faults).

Most software reuse libraries are source-code libraries so it may also be useful to store in the library

- the design of the component
- test-data and results for the component.

The specification may also include the size and complexity of the item if reus measurements require these values.

3.6 CONCLUSION

Measurement of reuse in software production is similar to measurement on new software production. In addition, measurements are required to manage the library of reused components: the impact of this requirement on the structure of the library deserves some consideration. The precepts and principles that apply to software measurement in general apply also to measurements on the reuse of software: a clear idea of purpose is essential before making measurements; one should also know how one intends to analyse the data before making the measurements; measurement theory applies; as always, care should be taken over data collection both to minimise the cost and motivate the staff concerned.

APPENDIX I

An illustrative example of software measurement

This small example is about software maintenance which is one form of reuse: the results are typical of what happens when one makes measurements on software production. Like 'structured' programming software reuse is one of those things that seems so obviously a 'good thing' that apparently few measurements have been made and fewer published. The following data was adapted from the Esprit 'METKIT' Project (METKIT Task 2.4, H. M. Sneed and A. A. Kaposi, 'A case study of the relationship between maintainability metrics and maintenance effects') to illustrate some typical measurements on a software production 'process'.

The purpose of the measurements was to evaluate the effect of source-code structure on the cost of maintenance. An old COBOL-74 program (COBOLD) was restructured using a commercially available tool to produce a new structured program of identical function (COBNEW). Another new structured version of identical function was handwritten using COBOL-85 (COB2NEW). Each of three programmers made three modifications to each program. The three programmers had three to eight years experience of COBOL but only one had experience of the particular application. The three modifications were typical of corrections, alterations and enhancements to program function. For each of the 27 (3 x 3 x 3) programs produced, the time to make each modification was measured and three other measurements were made, viz. the percentage increase in the number of lines of code, the number of errors in the program and the graph-complexity of the code, as measured by a static analyser tool. Only the average graph complexity was reported. Some data from the study are given in Table 1.

Programmer	A(mins) a	b	c	B(% loc.) a	b	c	C(errors) a	b	c	D
Corrections										
COBOLD	45	30	43	1.8.	3.4	2.9	0	1	1	29
COBNEW	90	36	75	7.7	3.5	2.2	0	1	0	26
COB2NEW	50	24	45	6.0	2.7	2.9	0	1	0	21
Alterations										
COBOLD	78	45	70	18.5	10.1	18.4	1	1	1	34
COBNEW	50	55	54	12.2	11.3	12.2	0	1	1	26
COB2NEW	35	37	40	12.0	12.2	6.6	0	1	1	23
Enhancements										
COBOLD	127	65	85	30.7	15.2	30.7	1	3	2	37
COBNEW	92	58	70	24.5	12.0	17.4	1	1	2	31
COB2NEW	74	47	70	26.8	13.5	18.7	1	1	1	24

Table 1   Measurements from a METKIT software maintenance study

Column A records the measured values of the time taken. Column B shows the percentage change in program size. The data in Column C records the number of errors in the changed programs. The average graph-complexity for each program is given in Column D.

It is not appropriate here to give an analysis of statistical significance of the differences between the three programs but Table 2 shows some totals which include the data from all three programmers.

	Total time	Average %inc. loc.	Total errors	Increase of complexity
COBOLD	588	25.5	11	8
COBNEW	580	17.3	7	5
COB2NEW	422	19.6	6	3

Table 2

It appears that comparing COBOLD with COB2NEW

- more time was required to modify COBOLD;
- the modifications produced a larger % increase in the size of COBOLD;
- fewer errors were made in modifying COB2NEW;
- the changes produced a smaller increase in the complexity of COB2NEW.

A similar analysis comparing programmers 'a' and 'b' indicates that programmer 'b' used less time, produced less code and made more errors than programmer 'a'(see Table 3).

Programmer	Total time	Average %inc. loc.	Total errors
a	641	27.3	4
b	397	13.6	11
c	552	22.3	9

Table 3 An analysis of the data by programmer

A much more comprehensive, well designed and well analysed investigation of the same subject was reported by V. R. Gibson and J. A. Senn in 'System Structure and Software Maintenance Performance', Comm. ACM., Vol 32, March 1989, pp. 347-358.

4 Migrating Towards Software Reuse

Ian Reekie
Instrumatic

4.1 INTRODUCTION

The most common question posed by software managers faced with the task of maintaining their competitive edge in a constantly evolving and expanding Industry is, *"How can we reduce our time to market and development costs, while increasing quality?"* While there is no simple answer to this question, increased software reuse, combined with the use of formal and semi-formal methods for software development, is one approach to realise these goals.

The adoption of a formal method for software development is now seen as probably the singular most immediate route to protect future software investment and increase reuseability. Automated in CASE tool environments these methods significantly increase and enhance productivity while improving the quality, reuseability and maintainability of code. There is still however significant room for improvement.

While methods and CASE tools promise to improve the reuseability of new software systems there still exists a vast untapped wealth of existing software applications that were developed many years prior to the adoption of methods and tools. The challenge is to unlock these resources and simplify the task of reusing and maintaining existing software.

This paper will review some of the methods and tools that have evolved to allow existing non-method/CASE developed software to be reused, together with proposing a practical approach to enhancing the reuseability of method/CASE developed systems.

4.1.1 The problems of migrating towards reuse

Systems developed without methods

The problems hindering the reuse of old software systems is often one of simple comprehension. Often these systems have evolved over many years with documentation slowly lagging behind the source code. As these systems have grown in size and complexity so has the problem of comprehension. To further complicate matters it is not uncommon to face these situations without the original designers, who by now have usually moved on to new and more challenging tasks.

Faced with this scenario even simple modifications are time consuming and tedious, while estimating the cost of product enhancements becomes a gamble, based on product intuition. What is required is a range of methods and tools that address the problem of comprehending existing systems to simplify the tasks of maintenance and software reuse.

Systems developed with methods and CASE
Software systems developed with the aid of methods and CASE support offer a far greater potential for future reuse. Usually designed using methods including event response modelling or object oriented techniques, they enable concise and simple identification of reusable elements, or subsystems within a system. While this certainly enhances reuse there are still many bridges to be crossed before total reuseability falls within our grasp. One such bridge crosses the divide between software design and software implementation and test.

Currently the transition from design to tested code, while being one of the most critical phases of software development, is often the least automated and most prone to human creativity. While the code generated may fulfil its designed function the possibility exists that it may be of low quality, contain hidden features and be extremely difficult to test. What is required are methods and tools support to quantify the quality and testability of generated code to further enhance the reuseability of software elements

4.1.2 Solving the problem of reuse

Two application areas have so far been discussed where the benefits of software reuse can be applied:

- reusing existing designs developed without the benefits of methods or tools;
- improving the current method/CASE based design process for reuseability.

Methods and tools will now be discussed that offer a practical opportunity to improve software reuse in each of these application areas. All that now remains is to examine in more detail some of the techniques and tools that are available to facilitate migration towards software reuse.

4.2 REUSING EXISTING DESIGNS

Reusing software from an existing system poses a number of problems, often the most fundamental being simple comprehension of the software's operation and internal architecture. Comprehending a software system usually begins at the most obvious sources of information such as documentation and listings. Without the information technology (IT) benefits of CASE it is not uncommon to encounter systems whose documentation suffers from one or more of the following characteristics; being incomplete, unwieldy, or out of date.

Without concise information software reuse becomes unrealisable. To overcome this techniques have been developed enabling direct comprehension of the software from the source code. Fundamental to this technique is the assumption that the software contains sufficient recoverable information. One such approach will now be considered that can be split into the following processes:

1. Identification of all fundamental building blocks within a system. These may be modules, tasks, routines or exceptions. Included in this step is the identification of the quality and complexity of each element

2. Identification of the architectural design relationship between all the fundamental building blocks of a system.

3. Requirements traceability, determining the elements or subsystems that implement a system function.

Each of these stages will now be considered in more detail.

4.2.1 Identifying fundamental building blocks

With the majority of current 3rd generation languages (C, COBOL, ADA,) the

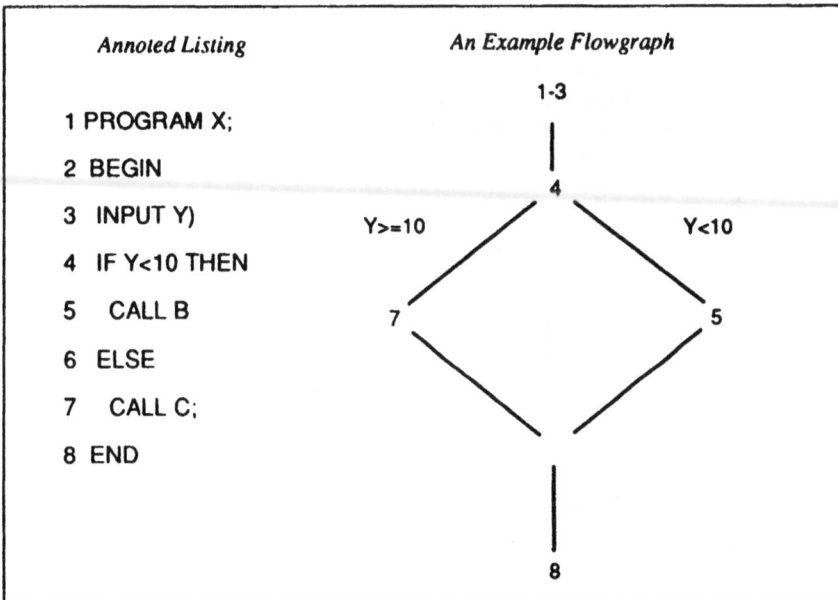

Figure 4.1

task of identifying the fundamental building blocks from source code is a relatively straight forward procedure, each building block mapping directly on to a program element such as a function, module or procedure.

While this approach does identify the original programmer's basis building blocks, it provides no guide to the complexity or quality of each element. For example, one programmer may have implemented a system with a hundred easy to comprehend modules, while his colleague may have implemented a similar design with forty, never to be understood, personalised works of art.

To enable code to be reused it is therefore necessary to have a method of measuring the quality and complexity of the basic building blocks in order to determine their suitability for reuse and maintainability. One of the methods available to measure these attributes is through the application of suitable metrics.

4.2.1.1 Using metrics to measure quality

Software metrics present one effective method of determining the quality and complexity of unit elements of software. To be effective metrics should include a series of characteristics:

- They should be meaningful.
- They should provide direct correlation potential errors.
- They should offer direct correlation with complexity and testing effort.
- They should be language independent.
- They should be automated.

One set of metrics that is an industry standard and has been partially adopted by the USA National Bureau of Standards are the McCabe metrics. Offering a range of measurements they are the basis for a range of methods and tools available from McCabe and Associates.

4.2.1.2 The McCabe metrics

The McCabe metrics are derived through the static analysis of program source code. The method first involves representing the code that is to be analysed in the form of a graphical flow graph. (see Figure 4.1)

The McCabe metrics are then extracted from the flow graphs using graphical analysis techniques resulting in the following metrics.

Cyclomatic complexity metric: Now a National Bureau of Standards adopted metric, the cyclomatic complexity metric is a direct measure of complexity. It is defined as, 'The basis number of paths through a program that cover every code statement and exercise all branch conditions'. By calculating the basis number of paths through a program element, the metric provides a direct indication of how difficult an element will be to comprehend, (see Figure 4.2). The metric also provides a direct measure of the testing effort required for each element. The basis paths directly correspond to the minimum tests required to execute all statements and exercise all branch conditions.

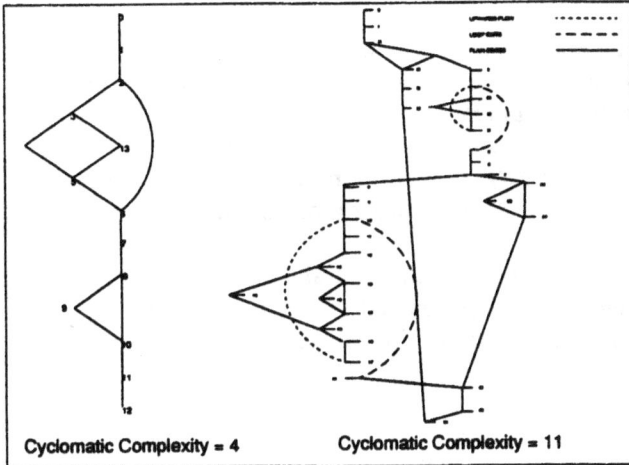

Figure 4.2 Example flowgraphs

Essential complexity: The essential metric provides a measure of the modularity of a program element and provides an intuitive guide to the effort required to comprehend or maintain a program element.

The essential metric is first derived by representing the program element as a control flow graph. This graph is then subject to a reduction process that removes all modular elements. What is then left represents the non-modular component of the program, the complexity of this is measured using the cyclomatic measure to produce a value for the essential complexity.

Design complexity: The McCabe cyclomatic and essential metrics measure the quality and complexity of a program element in isolation. In order to gain a measure of the total complexity it is necessary to consider its interactions with other elements.

The McCabe design complexity metric provides a direct measure of the interaction between a program element and the other elements in the system. It is derived from determining the number of paths though the program element that would have to be traversed to exercise all interactions with other modules. Thus for a program element having five total basis paths calling two other program elements, the design complexity would be three. Two paths to test each of the calls and one path to test a non-calling path.

The design complexity, while providing a measure of the interaction between elements, also directly responds to the basis set of integration test paths for a specified element.

4.2.1.3 Applying the McCabe Metrics

The McCabe metrics provide an immediate method for gauging the suitability of program elements for reuse.

> *The cyclomatic complexity*: provides a guide to the overall complexity of a module, being derived from the number of basis test paths
> *The essential complexity*: provides a measure of the modularity of a program element, a factor proportionately related to maintenance effort.
> *Design complexity*: provides a measure of a program element's interaction within a system. A factor proportionately related to reuseability.

Applied to software systems the McCabe metrics provide an immediate method of judging the cost timescales and implications of reusing, enhancing or maintaining program elements within a system. Below are some of the typical inferences made after application of the McCabe metrics:

High cyclomatic complexity: A number of case studies have been undertaken to determine the point at which programs become difficult to comprehend and reuse. While there are no absolute guidelines, a cyclomatic complexity of ten is usually taken as a threshold. It is not unusual to encounter existing systems with cyclomatic complexity of many hundreds. In the event of high complexities elements should be reviewed to determine if they can be split into smaller, more maintainable elements, assuming the essential complexity metric indicates this is possible.

High essential complexity: A high essential measurement for a program indicates that the module will probably be difficult to comprehend. Modifications within such a module will probably indirectly impact on areas not requiring change, leading to the familiar scenario in older systems of one error introduced for one removed. Thus if changes are required to modules with high essential complexities, redesign is often the most prudent course of action.

High design complexities: The design complexity metric is a direct guide to the integration complexity of a program element. Elements exhibiting high design complexities will require the appropriate number of integration tests to ensure comprehensive test coverage. For program elements with low cyclomatic and essential complexity metrics, high design complexities can be easily comprehended. However, with high cyclomatic and essential complexities the program element would probably require reduction into more realisable elements.

4.2.1.4 Using metrics to detect duplicate and redundant code

One problem often facing the reuse and maintenance of code within large systems is the identification of duplicate or redundant code generated during previous maintenance activities.

By using selected metrics individual elements that are suspected of duplication can have their metric fingerprint calculated. Scanning for similar fingerprints can then reveal modules with similar structures. The process of comparing metric fingerprints is only a coarse method of comparison. To achieve a closer match it is necessary to subsequently compare the detail of the individual programs. If large numbers of modules are concerned this can be tedious and error prone.

To automate this procedure McCabe and Associates have developed an automated tool called Codebreaker. Utilising the McCabe metrics, the tool calculates the basis test paths through the reference element suspected of duplication. By comparing test paths from the reference element, with those calculated for all the elements identified during the coarse scan, the tool identifies any matches. By altering the resolution of the tool, modules with similar structures or subsystems can be identified allowing the system to be reduced with direct impact on reuseability and maintainability.

4.2.2 Design architecture recovery

The McCabe metrics provide a guide to the maintainability and reuseability of the basis elements such as modules or procedures within a software system. Once all fundamental elements have been examined the next step on the path to comprehending the system involves the extraction of the overall design architecture (assuming there is one to extract).

For the majority of existing programming languages the design architecture can be recovered at the lowest level from analysing the source code directly and interpreting the calling hierarchy (see Figure 4.3).

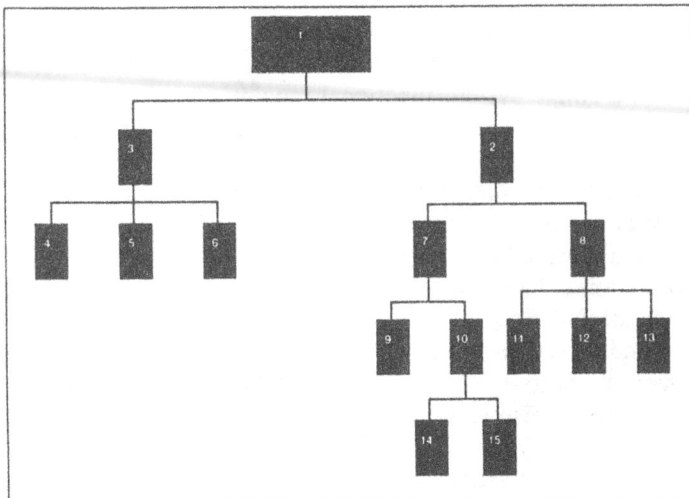

Figure 4.3 A typical design architecture (Program elements 1,15)

Recovery of the calling architecture provides one part of the design jigsaw. Further information can be derived from analysing the data usage of the individual program elements. This information can be correlated with the program calling hierarchy to enable structure charts to be displayed with their appropriate data couples, together with the full population of a data dictionary.

The process of design recovery is currently attracting significant research and development from a number of the key CASE tool vendors. The goal being to provide full design recovery from source code to a CASE design environment thereby allowing further forward engineering to be undertaken utilising the full benefits of CASE. This design and development is bearing fruit with highly integrated forward and reverse engineering CASE environments becoming available for languages including ADA and C.

4.2.3 Requirements traceability

A number of the techniques have been considered to:

- Analyse an existing program to determine the quality of individual program elements.
- Recover the design architecture and data dictionaries.

Solving the next stage of the reuse puzzle involves tracing which elements or subsystems within the software system are responsible for requirement level features/functions. This task is often the most complex and resource intensive stage of reusing or maintaining an existing program.

In the ideal world requirements may map clearly onto the design structure and code, allowing the implications of modifications to requirements to be easily gauged. However, in older systems direct mapping is usually not possible and it may take extensive effort to locate the appropriate code, elements, or subsystems that implement a single system functional requirement. One method of tracing requirements is to use dynamic requirements traceability techniques.

4.2.3.1 Dynamic requirements traceability techniques
The technique of dynamic requirements involves tracing the execution of a software system as a feature/function is exercised in real time. From this trace it is possible to determine the software elements within the system that are responsible for the feature/function.

McCabe and Associates have refined this technique and combined it with design recovery tools to produce a tool called Slice. Slice embodies the static design recovery mechanisms discussed so far, with the capability to execute and trace software systems dynamically.

Slice functions by performing a static analysis of the software system to determine the control logic of the individual program elements and design architecture. These are then used to build a complete picture of the systems software structure. While performing this analysis the tool also creates a copy of the executable software that contains software instrumentation at all branch or decision points. This subsequently allows the software to be traced.

The next stage of the Slice process involves the execution of the instrumented program under its usual operating conditions, exercising each function/requirement in turn that requires a trace. As each function is executed the instrumented version of the software generates a dynamic trace file with sufficient information to allow subsequent analysis to trace a function though the statically recovered software structure, (see Figure 4.4). This effectively provides an execution slice though the system.

The execution slice is a real time record of the code executed within a system to implement a feature/requirement. This information includes all elements that were touched by the slice, and provides resolution down to the individual lines of code that were executed.

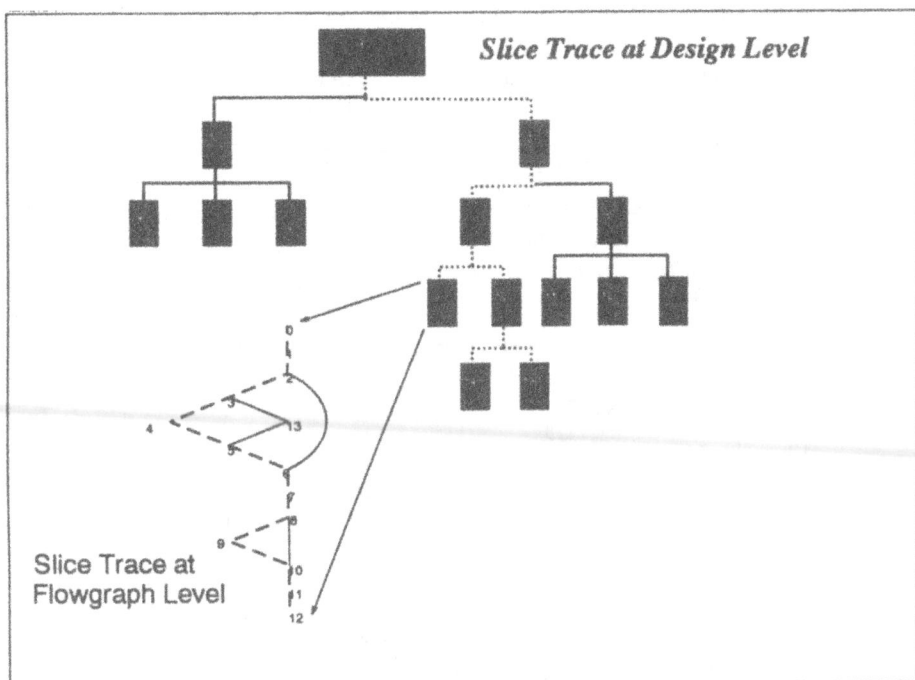

Slice Trace at Design Level

Slice Trace at Flowgraph Level

Figure 4.4 Slice

Slice provides a simple and effective method of tracing a function or requirement through a software system, enabling the impact of changes to be estimated and costed, while also simplifying the process of software reuse.

4.3 IMPROVING THE CURRENT METHOD/CASE BASED DESIGN PROCESS FOR REUSEABILITY

Current software systems developed, using CASE within the framework of methods, offer the promise of being easily reusable, their modular or object based architectures enabling concise and simple identification of reusable elements. This, combined with the high standard of documentation associated with CASE, greatly simplifies the task of reuse and maintenance. However there is still considerable room for improvement.

If the software development lifecycle is considered (see Figure 4.5), there is now considerable maturity in both methods and tools to support all phases of analysis and design. As these have evolved, following in their wake have come methods and tools to address the latter half of the software development life cycle, involving software generation and test. It is in this evolving portion of the software development life cycle that the most practical and immediate improvements can be made to increase reuse.

The latter half of the software life cycle has become known in some development circles as the 'BACK END' of the life cycle. It is in this area that we cross from design to implementation, often with the bare minimum of automated tools. During this transition, without the comprehensive checking provided by CASE, it is common for errors to be injected and designs implemented with no thought to their quality, maintainability or testability. To enhance this process, a new range of tools are beginning to present themselves under the guise of Back End CASE tools, also known as Computer Aided Software Test or CAST. These CAST tools provide comprehensive checks on the quality and testability of the generated code, while automating test generation and verification.

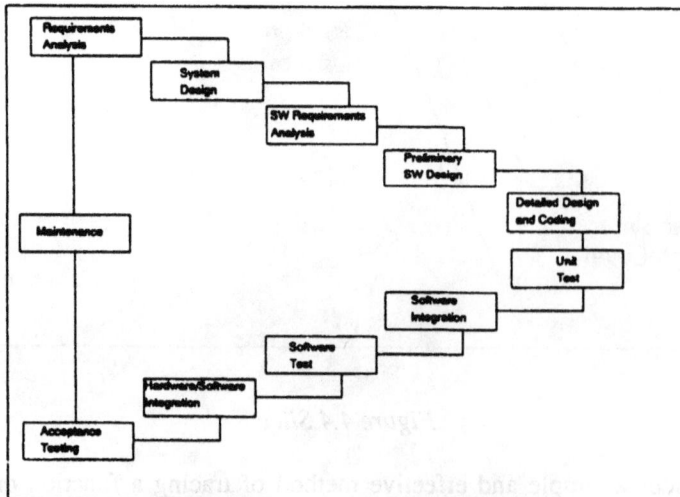

Figure 4.5 The software life cycle

4.3.1 Measuring software quality and testability

The most effective method of measuring software quality and testability is through the application of metrics, such as the McCabe metrics. These metrics provide an immediate guide to the quality of code being generated and ensure that the transition from design to code produces elements that are easy to comprehend and maintain. For further information please refer to the earlier section on metrics.

4.3.2 Comprehensive test generation and verification

Comprehensive test generation and verification is an essential step in the process of producing reusable code. It is the final guarantee that the reusable element conforms to its design requirements and fulfils its desired function. While testing is essential to reuseability it is often the least automated and measurable development activity.

To correctly test a software element or subsystem it is important that the tests include two classes of software tests:

- Black box tests, to ensure that the reusable code fulfils its design requirements.
- White box tests, to ensure that the logical implementation of the program element is correct, with no added functionality or unwanted features.

Black box tests have traditionally been derived directly from the system requirements documentation, a tedious and time consuming task that was highly prone to error. More recently tools have been introduced from CASE manufacturers, notably Cadre Technologies, that enable requirements encapsulated in a CASE analysis or design model to be directly translated into test cases. This promises to radically improve the quality and reliability of requirements generated tests by effectively automating a time consuming task.

While Black box tests aim to test an element against the requirements, white box tests are more concerned with the hidden implementation or logic. These tools provide important information on the number of possible routes or decision paths though an element. McCabe and Associates are proving to be a key innovator in this area: building on their innovative metrics the tools automate the generation of unit or integration level white box tests directly from the source code. Test verification may also be achieved using dynamic verification tools that ensure complete testing.

4.4 CONCLUSION

The migration path to software reuse is not a simple one, and there are no tools that deliver the panacea of instant software reuse at the press of a button, (at least not yet!). Instead, the path to reuse is a complex process involving a range of methods and tools to enable existing systems to be comprehended and reused, while improving existing development techniques to support future reuse.

This paper has reviewed a range of methods and tools to support and enhance the process of reuse in two application areas:

- Reusing existing designs developed without the benefits of methods or tools.
- Improving the reuseability of systems developed with methods and CASE.

For existing designs, a range of innovative software metrics, methods, and tools has been discussed that enables software systems to be analysed directly from source code, thereby providing information on attributes including software quality, testability, and maintainability. Further analysis techniques have been reviewed that facilitate requirements traceability and design recovery, enabling systems to be comprehended for reuse and maintenance even when documentation is unavailable.

While enabling the reuse of existing software systems, the metrics and techniques reviewed may also be applied to systems developed with methods and CASE. By measuring code as it is generated from the design specifications for quality and testability, appropriate thresholds may be imposed thereby enhancing the future reuse of these software systems.

REFERENCES

Gibson, Virginia R. and Senn, James A. System Structure and Software Maintenance Performance, Communications of the ACM, March 1989, **32**.3, pp. 347-358.

McCabe, Thomas J. and Butler, Charles W. Design Complexity Measurement and Testing, Communications of the ACM, **32**, December 1989.

McCabe, Thomas J. Reverse Engineering, Reusability, Redundancy: The Connection. American Programmer, October 1990.

Structured Testing: A Software Testing Methodology Using the Cyclomatic Complexity Metric, National Bureau of Standards Special Publication No. 500-99 December 1982.

5 The Component Management Road to Reuse

Malcolm Fowles
DCE Info

5.1 INTRODUCTION

Software reuse does not happen by magic. Although a particular technology may provide more reusability, this is merely potential until you turn it into actual reuse. To do this, you must apply the technology in a process which makes the reuse happen and maximises what you get out of it. In other words, you must manage it. Success stories from Japan and the USA suggest that you can reuse anything as long as you organise properly, which would make reuse purely a management issue.

However, common sense also suggests that you can waste good managers fighting technology that is bad for reuse. The software industry has always looked enviously at others using component technology to reuse the same design over and over again, in a flexible, modular assembly process which confers great productivity and quality benefits. It is time to reappraise the applicability of this approach to software in the light of new developments such as object technology.

5.2 MANAGED REUSE

5.2.1 General

Reuse is meaningless in respect of a single product, being the feedback of material from one product to another. When the feedback is explicit rather than left to chance, such that effort is put into engineering reusable material and all new initiatives look first to see what the material can provide, then we can say the reuse is organised and the material is an asset. When in addition we set goals for the feedback, and monitor the reuse that is achieved, and act to improve it, then and only then can we claim to be managing reuse.

5.2.2 Planning

Both planning and monitoring focus on goals for reuse and their measurement.

Another paper (Mole) looks specifically at metrics. Here are some metrics which help to explore the effects and limits of different approaches to reuse, and to plan improvements:

(a) Reuse efficiency. The fraction of the cost of an average component that can be saved by reuse. Reuse can never eliminate all costs, such as integration testing.

(b) Reusability overheads. The fraction to be added to the cost of an average component to engineer it for reuse. This would include apportioned investment in tools and skills, management overhead, and extra design effort.

(c) Level of reuse. The fraction of an average product that is made of reused material. The characteristic of software is its variation, which will always demand some new material.

(d) Occasions for reuse. How often is the average piece of reusable material used?

(e) Impact of change, a sub-goal of reuse efficiency. The average number of components affected by an individual change. Reuse is never a simple choice between to use or not to use. When material is 'almost' right, we can customise it. With conventional software technology this is the rule rather than the exception.

The first four metrics are based on economic models of the effect of reuse on productivity and quality (Gaffney and Durek, 1989). Other models can be produced for other desirable effects. Models let us explore the implications of different levels of achievement. For example, we can see that quality improvement is a more effective short term objective because its benefit is mostly achieved with two or three occasions for reuse.

The basis of planning is the definition of *repeated* need for *common product characteristics*. Each identifiable need breaks the problem down. Each presents different opportunities for reuse and there are different influences and limits upon success. For example, the scale of repetition limits the goal of occasions for reuse. Some characteristics, such as a common user interface, might be needed many times in many products while others, such as the use of special hardware, might be rare.

It is useful to treat each such opportunity as a market. No matter how large the market, you still have to penetrate it, i.e. to ensure that people do actually reuse material to satisfy the repeated need. Another paper (Kruzela) describes a management structure to overcome several typical obstacles, but there will always be others specific to a situation, such as hardware variations which increase the costs of reuse.

To be rigorous you must also allow for planning errors. You will be far more certain about the need for some components than others (in a volatile business area, for example) and should allow for this risk when deciding priorities.

The other influence on reuse goals is of course the technology. How much can it save and how much of this depends on the quality of your work with it?

Different technologies will support different markets for components: an executable code library for user interfaces; template source code for transaction structures; CASE definitions for business rules.

Strategic planning would analyse the reuse efficiency and other parameters of technical alternatives. Such an analysis will show, for example, why object technology is particularly 'reusable'.

5.2.3 Organisation

Another paper (Kruzela) deals with the organisation of reuse. Whatever this may be, it will not be the familiar, project oriented, development shop where such responsibility as is formally recognised rests with a relatively minor technical function. As the level of reuse grows so the management of reused material becomes more important than original development. At this point the responsibility must have risen to the top and new management functions must be carrying it out. Otherwise it will fail.

5.2.4 Business process

The new organisation can be devised as an overlay on the redesigned business process of providing systems support. This is effectively a new systems development life cycle. It incorporates the reuse feedback loop and separates activities which make no sense to a single project. Here is a brief tour round the process.

(a) The first corporate activity is planning, such as that described earlier. It tries to maximise the benefits of reuse by anticipating enterprise-wide demand for software.

(b) Component design, or acquisition, actually produces the reusable material. No formal responsibility is placed on projects to do this – the overhead is accepted. Note the new outlook on re-engineering. It is technically easier and makes better economic sense to extract components from existing systems than to re-engineer them entirely.

(c) Reusable material is managed in an inventory to be as attractive as possible – easy to find, to understand (see the paper by Maiden) and to extract.

(d) The first response to actual demand for software is to create an initial configuration from the inventory. The obvious example is a first cut prototype, but it could equally be a specification made from corporate standards. Here the potential for reusable components from object oriented analysis and design (see the paper by Lovegrove) offers a much greater power to succeed than conventional models.

(e) As was pointed out earlier, no new software product will be made up entirely of reused components. The world moves on, and this is probably why the new product is wanted anyway! The next step is to meet the actual demand by evolution, i.e. by customising the initial material and adding more. As the level of reuse increases and prototyping, or code generation, tools get better, this stage will shrink and one of the objectives of reuse – responsive software production will be reached.

(f) One may assume that there is work outside the scope of the material in the inventory, seen as a separate delivery stage.

Component management must be a service to the production function. It cannot enforce reuse that is uneconomic nor be a delaying bottleneck (as are some central services, like data management, in too many organisations). There is no alternative but to save developers' time and money. A service level agreement will provide the necessary formality.

5.2.5 Tools

The shift in emphasis away from software origination has a profound effect on tool requirements. Tools to manage what already exists, such as lists and exceptions, bulk changes, impact analyses, and so on, become more important than drawing aids to create new models. If you need a new configuration there should be a tool to do it for you. Method support is needed at the enterprise level: an individual system is configured from pre-existing corporate analysis and design and evolved by prototyping thereafter.

5.3 PROSPECTS

5.3.1 Appraisal

How far away is this picture of component management from today's world of software development! We carry a huge investment in conventional technology where the impact of change is so great that we have to consider full system re-engineering. Most current tools and techniques, in which we have also invested heavily, positively encourage origination at the expense of reuse. Product planning is rudimentary. Design skills of the level needed to create components are little known and may be beyond the average capability. The individualistic, project oriented business process requires a thorough redesign and cultural upheaval. You would be forgiven for thinking that component management is a nice idea, but irrelevant.

In the average IT department that may be true. How many times will the average component be needed? There are possibly not enough occasions for reuse to justify the reusability overhead, especially if that involves changing the world outlined above. A few pioneers provide glowing exceptions, but most enterprises lack the imagination and the courage for such a change.

One class of software developer is less constrained. Product suppliers are driven by competition, which weeds out unfit technologies, unwanted features, uneconomic practices, etc. Their product planning is their marketing, and they have unlimited occasions for reuse: even two occasions gives them more expertise than any customer. Their potential return on investment justifies high expenditure on quality resources.

5.3.2 Trend

For a long time product suppliers have been confined by conventional technology to the concept of a replicated package that could be varied only at great expense to meet a customer's needs. Recently there has been a move to mass customisation founded upon component management. Suppliers like Clebern and CTP offer solutions that are as cheap as a package and as fitted as a well-prototyped bespoke system. They can sell direct to the business on price, time scale and quality, by-passing the IT department.

5.3.3 Issues

The reason why mass customisation has not yet swept this industry as it has others is the issue of interoperability, both with existing systems and with future ones. For as competition moves towards specialisation one supplier cannot offer a solution for everything. Here lies the need for industry standards.

The successful mass customisers of software rely on object technology, confirming its claim to greater reusability. Standards in this field are well advanced, and can be said to have leapfrogged conventional standards with the acceptance on 11th September, 1991 of the Object Management Group's 'Object Request Broker' standard. ORB compliant products will be able to use each other's services, leading to the concept of the best mix of applications in a field, using the best help and tutorial system in the world, via the best user interface on a particular platform, and so on.

5.4 CONCLUSION

Thus have the pieces fallen into place for component management to become the driving force of software development, but not in the way most of us imagined.

Component management can only work where, instead of manufacturing software in response to the arising of demand, we manufacture parts in anticipation of future demand. The burden of software's diversity moves from manufacture to assembly, which is why it needs technology such as object orientation with a low impact of change and a high level of reuse.

The only low risk way to anticipate demand is to influence it, so component management may be irrelevant where marketing cannot employ it. This view is supported by the most cursory glance at other industries. As Will Hutton (1991) said of mass customisers, the winners 'use the flexibility to completely rethink how a company serves its markets'.

The adoption of component management by software suppliers has already begun. The laws of economics make it only a matter of time before they replace the in-house IT department as the source of new business systems support. I see a striking parallel with the car industry: before component-based production concentrated manufacturing into large units able to afford the investment, virtually any back street garage could build you a car. The IT department is like a back street garage.

REFERENCES

Gaffney, J.E. and Durek, T.A. 1989, Software reuse – key to enhanced productivity: some quantitative models. *Information and Software Technology*, **31** (5), 258-67.

Hutton, W. 1991, New brooms sweep out production lines. *The Guardian*, (16th September), 11.

6 Software Reuse: State of the Art and Survey of Technical Approaches

Alistair Sutcliffe
City University

6.1 INTRODUCTION

This paper sets out to survey different approaches which have been applied to the problem of software reuse. First a historical perspective of reuse research and practice is outlined and the reasons for failure surveyed. Then a framework for evaluating various approaches is given by stating a set of problems which reuse must solve, both in technical and managerial terms. The approaches are then compared against this framework and their progress to date, and future prospects, are discussed. Before continuing, some clarification is necessary about the nature of reuse. In this paper I shall consider reuse at several levels of ambition, software code, designs, specification and even ideas and concepts. A second dimension is the span of reuse between domain or problem types; at one end there is reuse of components in similar domains, at the other end reuse may be possible between very different domains. Teasing apart the dimensions of the reuse problem is another objective of the paper.

6.2 REUSE – A HISTORICAL PERSPECTIVE

Software reuse has been effectively practised at the implementation level for many years in the form of code libraries, e.g. COBOL Data division libraries, FORTRAN modules from the National Algorithms Group. Early interest in reusability was targeted on program development and this achieved modest results in terms of small scale modular reuse (Freeman, 1987; Biggerstaff and Richter, 1987; Meyer, 1988). Program module reuse primarily focused on the 'black box' approach which assumes no internal knowledge is necessary. At the design level abstract data types, ADTs (Guttag, 1975), were envisaged as reusable components. By defining composite data structures and associated actions which fulfilled some functional purpose, it was hoped that libraries could be created to facilitate reuse. However, early hopes have been frustrated; ADT libraries do exist but their impact on reuse has been disappointing.

One motivation for the development of structured methods was to enhance reuse by design of modular software with high cohesion and low coupling

(Yourdon and Constantine, 1977). The aim was to produce independent software modules (low coupling) which were readily identified according to their processing goal (functional cohesion) and hence facilitate reuse. Unfortunately this dream has proved to be elusive. Identification of cohesive modules has not proved easy and neither has construction of designs with low coupling.

More recently object oriented programming languages (e.g. Smalltalk) and associated design approaches have been developed to enable reuse. Object orientation (O-O) enhanced the modular approach by addition of classification hierarchies and reuse by specialisation, or inheritance, of properties of generic objects. However, methodical guidance and tool support for object oriented reuse are still absent or inadequate. The O-O approach is also limited because as the size and functionality of module/objects increases, knowledge of the internal specification becomes necessary so software developers can understand how modules should be applied. This limitation has led to concepts of 'white box' reusability such as transformation systems and reusable patterns (Biggerstaff and Richter, 1987) in which all parts of reusable components are available for customisation in a new application.

A few reuse success stories have been reported but these are exceptions rather than the rule. In particular practice in Japan shows promise where statistics in the range of 15% effort saving and 20-30 % module reuse have been quoted. However, in spite of considerable research effort and development interest, progress to date on software reuse has been disappointing. The following section reviews some reasons why progress has been slow.

6.3 BARRIERS TO PROGRESS

Barriers can be either technical or social in nature. Technical barriers include problem of finding and matching reusable components to a new application context; the design problem of composing a system from reused components; and the impact of more general software engineering problems (e.g. design for reliability and quality) on the reuse area. Technical issues will be covered in more depth when the evaluation framework is introduced.

Social issues were ignored for many years, although more recently these problems have been realised and some authors now see reuse as primarily a social problem (Frakes ,1991). A selection of social issues follows:

(a) Libraries and Repositories
Reuse will only take off when there are sufficiently large libraries of components for designers to use. Currently reuse libraries have hundreds of components, if that. Libraries of thousands are required. This is a re-run of the Teletext problem; only when a critical mass of information was available on the network-library did teletext take off, as in the Mintel service in France. With the spread of CASE technology and certain standards

for repositories (vis A/D cycle), the library problem may become more solvable.

(b) The Human Factor

Design, even software design, can be a creative activity. People are reticent to reuse someone-else's design for a variety of reasons. They may mistrust reusable components, they have to understand a reusable component which takes effort anyway, and finally people may prefer to design from scratch because they retain control over design quality. Another reason is plain habit. People like doing things the way they have always done them, and changing an operational culture is difficult.

(c) Management

Few incentives are ever given by organisations to overcome the human factor. Management incentives are needed to challenge the 'not invented here' syndrome. New working practices for design by reuse, and design for reuse should be encouraged with incentives. Few organisations have this management philosophy.

(d) Legal

If software reuse becomes common place it will change the nature of software from bespoke items to standard products. Reusable software may therefore become controlled by trades description and standards legislation, especially when software is reused in safety critical applications. Defining legal responsibility for failure will be a lawyer's paradise as they debate who is responsible between the first author, the reuser, modifier, end users and so on.

In conclusion, there are many barriers to successful reuse, many of which have been underestimated in the past (Walton, 1991). Future success is likely to depend on solving an amalgam of technical and social problems.

6.4 TECHNICAL PROBLEMS AND AN EVALUATION FRAMEWORK

Reuse can be happen at a variety of levels and several different interpretation have been reported. For instance reuse has a dimension which maps approximately to the system development life cycle:

Implementation level reuse: reuse of code modules, already an established practice.
Design reuse: documents describing designed components are the immediate object of reuse. The design may then be realised by tracing to a code-module library to find the implementation of the design or by automatic

code generation from a design document.

Specification reuse: this follows the same pattern as design reuse, although specifications are conceptually further from an implementation, hence more work has to be performed, either manually or automatically, to turn a specification into code. Alternatively, the corresponding code module which implemented a solution to a certain problem specification may be traced, but some modification may be necessary.

Knowledge-concept reuse: in this case it is understanding of the nature of the problem itself which is reused and possibly knowledge of previous solutions to similar problems. Such knowledge may of course be held within specifications and thereby traced to code modules.

A semi-orthogonal dimension is the span of reuse from very similar applications within one domain to dissimilar applications which share the same logical properties even though they appear in apparently different domains. Inter-domain reuse of problem types is only realisable at the higher levels of specification and concept reuse.

Another variant on the reuse story is to reuse the process rather than the product. The decision-making history of developing a particular application contains a wealth of potential reusable information about human reasoning for certain problem types. This could be reused to help people solve similar problems rather than re-inventing the wheel each time. Reuse of design traces or rationales also tends to be related to specification and concept reuse, although not exclusively so. The dimensions of reuse are summarised in Figure 6.1.

As reuse progress up the life cycle dimension, the difficulty and hence cost of solving the problem increases. On the other hand the potential payoff increases too because requirements engineering is acknowledged to be the most error-prone and therefore ultimately costly part of the life cycle. Reuse of ideas, concepts and specifications in requirements analysis could help improve targeting functionality to users' needs and understanding of applications by software developers.

Targeting raises the compatibility problem. Reuse at any level of ambition has to achieve a match between the reusable components and a new target context.

Compatibility divides into dimensions of coverage and targeting.

- The coverage of components may be broad, i.e. they could fulfil several different processing goals. In this case reuse in a restricted domain may mean throwing away some of the component's excess functionality. Alternatively, narrow coverage means specialisation is easy in a small number of target domains.

- Targeting expresses the degree of matching between the target application and reusable components. This will affect the ease of specialisation of a new component, e.g. how well the detail required by the new domain fits into the component's properties.

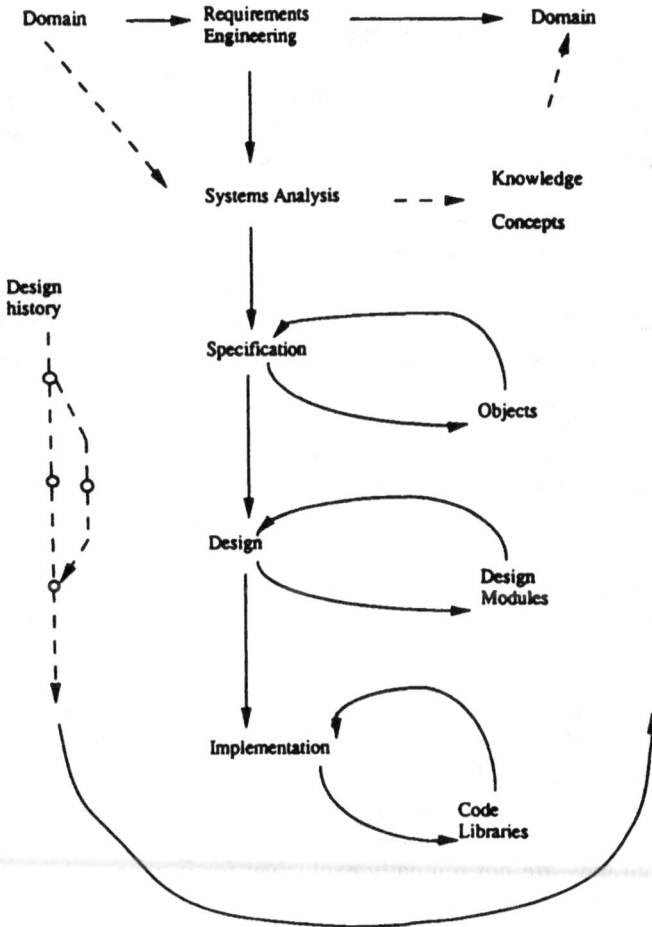

Figure 6.1 Perspective of software reuse

To achieve successful reuse four fundamental problems have to be solved:

(i) The Search and Retrieval Problem
Simple one word functional descriptions may be sufficient to extract any suitable candidates from libraries of a few hundred small unrelated routines. However, if we attempted to merge such libraries into a Common European Library of over 100,000 items, they would become unusable. Similarly an organisation with just 20 major applications built up over some years, consisting of say 100 sub-systems, is unlikely to have anyone who can easily match the requirements for a new application against

them. This problem requires a technical solution for searching repositories or component libraries.

(ii) Component Matching and Assessment
This is linked to the first problem but extends it by introducing goodness of fit. Retrieving any component necessitates some description of the target context. Retrieved components can then be matched and assessed for their potential suitability with respect to the new context.

(iii) Component Understanding
Reuse entails understanding the functionality of a component, even if it has a black box design and no knowledge of its internals is required. Unless designers are reusing components which are already very familiar, explanation and facilities to help component understanding are necessary.

(iv) Customisation and/or Composition
Once one or more components have been retrieved and understood a new system has to be constructed, either by tailoring existing components or by composing the application from building blocks. Although this process will require existing tools such as CASE-tool diagram editors etc. further support is required to help the system developer reuse components in a new context to make sure that errors are not introduced into the design during composition/customisation.

These four problems require technical solutions and tool support to enable reuse in all but trivial environments. This framework of technical issues and reuse approach is now applied to various approaches to reuse. These have been categorised as: Object-oriented, Formal methods, Domain Analysis and Application Frames. These categories represent the major themes of reuse research although they are not mutually exclusive; for instance, object oriented approaches may also involve formalism.

6.5 REVIEW OF APPROACHES TO REUSABILITY

6.5.1 Object oriented approaches

Reuse in O-O development is achieved either by inheritance and specialisation of generic objects (Meyer, 1988) or by using more specific objects as building blocks in applications (Booch, 1986). It is notable that many methods claiming to be object oriented (Ballin, 1989; Shaler and Mellor, 1988) ignore this critical aspect of object based development.

While the standard qualities of object orientation, e.g. modelling composites of data and activity, encapsulation, and classes can be found in many existing methods (Cook, 1986; Meyer, 1988), the reuse is not explicitly supported. Object oriented methods need two properties for successful support of reuse:

- Object oriented models should embody classification and inheritance mechanisms.
- Object models should be abstract and not contain low level application detail. Reuse should be facilitated by creating generic objects for subsequent specialisation in new applications.

Inheritance can be simple (i.e. properties are only inherited in strict hierarchical order from one parent) or multiple, in which case attributes or services (really algorithms but also called methods in O-O literature) may be inherited from different parents possibly in quite separate classes. Multiple, or polymorphic, inheritance is more flexible but it pays a penalty in terms of control and maintenance. When a service is inherited several times into many lower level objects tracing any changes may become a considerable problem.

Moreover the service may not work as intended in different specialised objects. Controlling multiple inheritance therefore requires tool support to trace the specialisation ancestry as well as to assess goodness of fit with different types of objects.

O-O methods should guide the developer in the practice of reuse, especially in the description of classes, and the processes of generalisation and specialisation. In this regard nearly all methods are sadly lacking. Even if abstraction is advocated no procedure is described for the reuse of generic objects in new domains. In addition inheritance mechanisms are often only partially specified so the process of specialisation is not clear.

Object orientation can be divided into the programming and system development. Early methods (Booch, 1986; Cook, 1986) were mainly a list of desiderata of the properties an object oriented program specification should possess, coupled with procedures for implementation in a suitable O-O language (e.g. Smalltalk, or the less suitable but more fashionable Ada). Numerous papers have been produced concerning different aspects of O-O development techniques and environments, e.g. inheritance, polymorphism, dynamic binding and so on. However, the endeavours of the O-O programming community have resulted in little effective reuse beyond small libraries usually focused on narrow problem domains.

More recently the structured methods community has realised the potential of object orientation. The first general treatment of object orientation from a methodological as well as programming viewpoint was Meyer (1988), and this work is still the authoritative reference. Meyer, however, falls short of giving methodical guidance about how to achieve reuse with O-O techniques. Since then a variety of O-O methods have been published.

Object oriented design methods

This group of methods are closer to the programming community and focus on design aspects.

Probably the best known method in this category is HOOD Hierarchical Object Oriented Design (Robinson et al. 1987). Objects are modelled in a hierarchical manner, with inheritance of properties and strong emphasis on the object interface specification and encapsulation. A system network of objects communicating by messages is created with control by event messages. HOOD uses Booch's conception of actor and server objects.

Object Oriented System Design (Wasserman et al. 1988), OOSD, provides a detailed notation for object classes, management of inheritance, interface description, and encapsulation.The system is modelled either as a sequentially executed hierarchy using the Yourdon structure chart notation, or as an asynchronous network of processes with monitors. No analysis advice is given so coverage of OOSD and HOOD is necessarily restricted to the design phase.

Object oriented analysis/design

A variety of methods have been described which address the analysis phase, some of which also attempt to provide some mapping to O-O design.

Shaler and Mellor's method, Object Oriented System Analysis, (Shaler and Mellor, 1988) gives heuristics for object identification and analysis but many of its recommendations are indistinguishable from Entity relationship modelling. The main criticism of OOSA is its lack of support for inheritance. Classes are supported but only inheritance of object properties is modelled. Inheritance of services is not considered and reuse is not explicitly supported.

Object Oriented Analysis (Coad and Yourdon, 1989) covers classification and inheritance. Three links between objects are supported: relationship connections, classification hierarchies and message passing. The method uses hierarchical inheritance and masking rather than multiple inheritance and specification of encapsulation and object interfaces is not as detailed as in OOSD or HOOD.

ObjectOry (Jakobsen, 1987) supports classification, encapsulation and inheritance. Abstraction is promoted by levels in design from higher level system views to lower block and component levels. Reuse is supported by component libraries and design transformations to real time languages. Guidance for analysis is less comprehensive.

For other Object Oriented Development methods see Wirfs-Brock et al. (1991), Booch (1991), and Rumbaugh et al. (1991).

Summary of O-O methods

O-O development methods still have to surmount considerable problems. The coverage of basic object oriented concepts such as inheritance is weak in some methods while the link to program design is poor in the O-O analytic methods. On the other hand O-O design methods give no guidance for O-O analysis and modelling.

More serious is the gap between methods and supporting CASE technology. Standard structured methods have only become accepted with the growth of CASE technology. The same will be true of O-O methods. Currently O-O CASE tools are hard to find. Superficial changes have been made by leading vendors and O-O method supplies so CASE diagrammers can support O-O type notations. But this is no solution. To help O-O CASE technology has to integrate O-O specification and design with libraries and repositories so that inheritance for reuse can be facilitated. This will take time and no method gives effective support to integrated development with reuse libraries, although O-OSD and HOOD can make some claims in this direction. Further details of O-O methods and a comparison with more traditional structured methods are given in Sutcliffe (1991a).

Support for O-O reuse

Retrieval support for objects has used either cut-down versions of classification schemes or offered no retrieval support at all, instead relying on editors and simple keyword based retrieval. Esprit projects have achieved some success in retrieval of documentation (MACS) and design for reuse (REDO) but reuse has been restricted to small scale applications, at the design level in domains with which the software engineer is familiar (e.g. DRAGON project, Wirsing and Hennicker, 1991). The only advance on the basic object oriented paradigm has been to provide design traces or notepads to record design histories for subsequent reuse. In spite of the considerable research and development activity on Reuse methodology and tool support in Esprit Projects (e.g. REBOOT, ITHACA projects, Funghi et al. 1991) results to date have been disappointing.

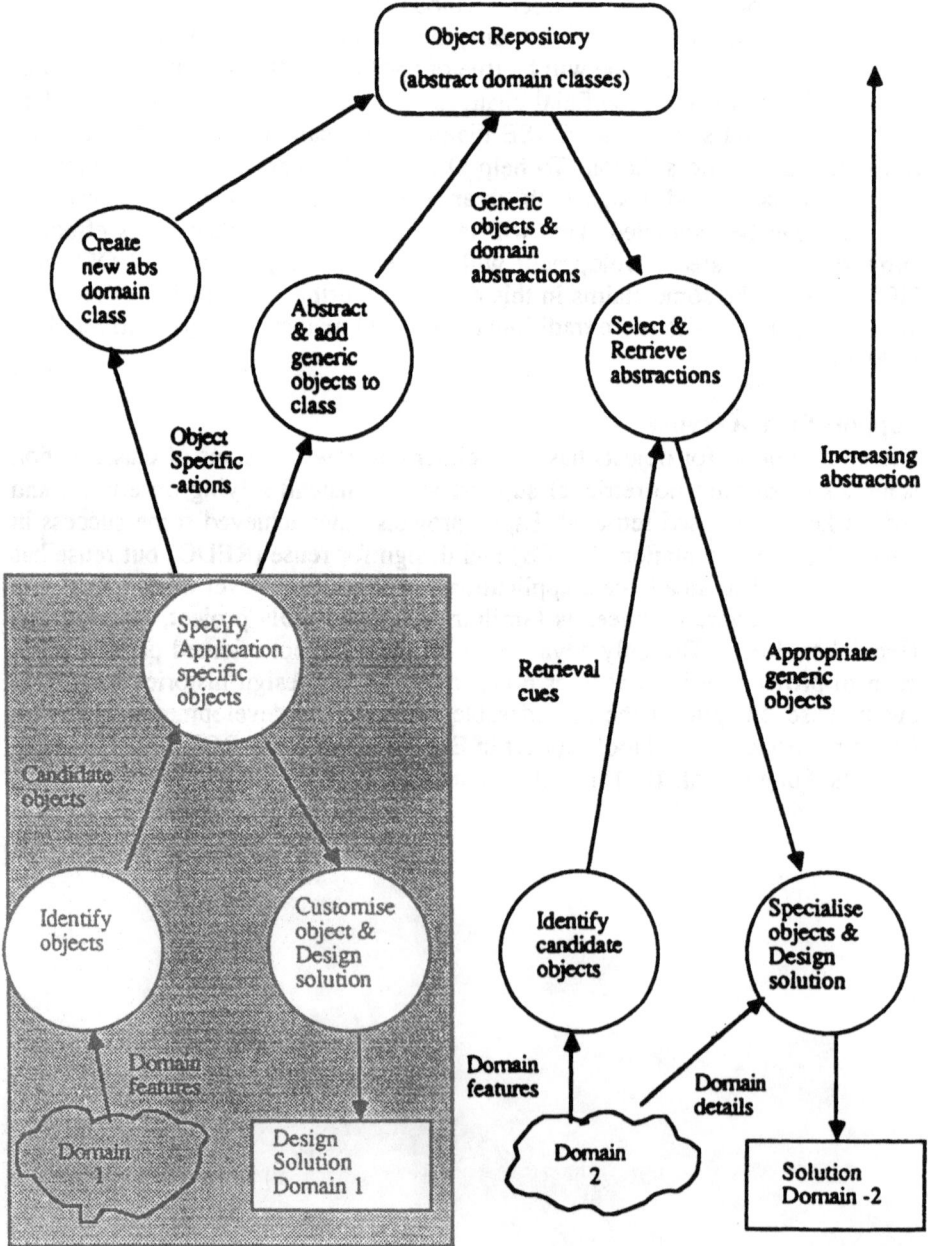

Figure 6.2 Model of reuse in object oriented system development

Another serious failing is the lack of method support for the reuse process. Little or no guidance is given for how reusable system development should be practised beyond bland assertions about specialisation of generic objects. The process of design for reuse, and indeed the problems inherent in specialisation are glossed over.

The process of abstraction is essential to successful development of generic objects. Abstraction is frequently used in its design sense as familiar computational structures such as stacks, trees, acyclic graphs, etc. In most cases the examples reported are small and simple, and the application is understood in detail. In the specification view, abstraction is the description of general properties of an application. There is, as yet, no consensus in the object oriented literature about just what a generic object should be and how much, and what sort of, detail it should contain. The most common approach appears to follow a data modelling approach, possibly modulated by the Jackson concept of active entities (Jackson, 1983; Sutcliffe; 1988). One view is to describe abstraction as the process for extracting the essential structural and processing components in an application by eliminating application specific detail. Another view is the description of common constructs from several related domains.

However, there is a boundary problem with both approaches: how related should the components be and how much detail should be discarded? The objective of abstraction for reuse is the description of the basic functional and structural aspects common to several application domains. Therefore, application specific detail has to be discarded as far as possible while preserving useful functionality. If a generic object is stripped of too much detail it will be of little interest for reuse. A model of the overall process reusable development is given in Figure 6.2.

First, application specific objects are specified and the system designed by composition of the these components. This is the development cycle implicit in most current O-O methods. Once an application specific specification has been created objects have to be 'generified' by abstraction. Candidate objects are retrieved for reuse and then specialised using further detail derived from analysis of the new domain. Current methods seem to ignore this process. Further details and a description of the problems inherent in trying to use OOA (Coad and Yourdon, 1989) for reuse can be found in Sutcliffe (1991b).

In conclusion the O-O approach shows some promise but it has failed to address the key problems in enabling reuse, specifically:

• Retrieval – few tools have been produced beyond simple keyword search facilities in editors and diagrams to show class hierarchies. These are hardly adequate for large scale environments and repositories.

• Matching and assessment – this problem is ignored, the assumption being that provision of the appropriate class hierarchy is all that is required. Little or no guidance is given about how to practice inheritance/specialisation with respect to a new application context.

- Understanding – O-O methods may claim to make an overall system more tractable by modularity and encapsulation, but no explicit help is given to help understanding and indeed encapsulation may hinder comprehension of an object's function and design rationale by hiding internal contents necessary for understanding its purpose.

- Customisation and composition – O-O may make a stronger claim here, as O-O designs should be composable by addition of more loosely coupled objects. Unfortunately little support for composition is given when inheritance becomes complex.

Overall the track record of object orientation is not good, a pessimistic observation considering one of the main rationales for the O-O approach in the first place was to facilitate reuse.

The final problem with this approach is that it does not address the mass of existing software. O-O can only deliver reuse if the reusable software itself was designed using O-O principles, tools and languages. Very little truly object oriented software currently exists.

6.5.2 Formal methods

Few, if any, formal methods have addressed reuse directly. The relevance of this approach comes from two directions; first, construction of libraries of formally defined components which may be reused (Moineau and Gaudel 1991), and secondly from formal methods themselves which may be applied to constructing reliable software which can facilitate reuse.

The most important contribution of formality has been in development of Abstract Data Types, which have now been built up into libraries (Uhl and Schmid, 1990). ADTs are design abstractions and can therefore help design level reuse. However, their abstract nature makes mapping ADTs to functional requirements difficult. What is missing is tool support for mapping (or matching) ADTs to early stages of the system development process. If description of a new application context could be used to match to ADTs held in a library then reuse by composition would be possible. Currently their use is restricted to the design level in problem domains which are well understood by the software engineer. The ADTs themselves must also be well understood to allow appropriate use. However this is not the whole problem. Construction of anything but a trivial system involves use of several ADTs. The higher level design of how these structures interact has not been solved, so although the components may be formally specified their interaction and overall behaviour may not be reliable.

This problem has focused attention on design objects and development by composition of generic design components. This is similar to the template driven approach (see section 5.4). Formality at this level, however, poses a considerable problem because the behaviour not only of components but also of their inter-communication has to be verified. This scaling up of complexity poses a problem for many formal languages (e.g. Z) although more recent process algebras (e.g. LOTOS) show some potential for solving this problem.

Formal methods may contribute to reuse by making the behaviour and the interface between components more deterministic and hence reliable. In design by composition Interconnection languages (Gougen, 1986) have been proposed by which the interface between modules is formally defined so a reliable system can then be composed of smaller well-defined components. If necessary the system could be composed of heterogeneous components and bridge modules developed to act as translators. Unfortunately interconnection languages have not been developed to deliver the promises of their proponents.

The final approach by which formal methods may contribute to reuse is in transformation programming and transformational development (Jacquart, 1991). If reliable procedures can be defined which change one design structure into another then design components could be reused by automatically changing them into another more suitable component. Design could proceed by a sequence of automatic transformations from specification to implemented code. At the design level this approach has been demonstrated, e.g. transforming data structures such as lattices into trees, queues and stacks, etc. However, the transformational approach runs into difficulties when problems are scaled up, and it cannot tackle specifications which do not readily translate (or transform) into well-known computational structures. Some progress is being made on this problem by Doerry et al. (1991) who take a layered view of successive transformations which are applied to different aspects (of views) of a specification.

Formal methods intersect with object oriented approaches in the formal specification of object interfaces, interconnections and communication. This has been a very active area of research and promises to increase the rigour of the O-O approach. While formalisation may make objects more reliable and could make their interface description more rigorous thus helping design by composition, formalisation is unlikely to solve the other reuse problems.

To summarise, formal methods may help reuse by providing libraries of reliable components and could improve design by composition. However formalisation does not solve the other reuse issues, e.g.

- Retrieval – formalisation may improve the basis of keyword and taxonomic approaches to retrieval (see section 6.5.3), but it gives no help to the definition of functionality and purpose of modules. Formalisation may indeed increase the 'description' gap between statements of need in a new context and the properties of reusable modules.
- Matching and assessment – formalisation could help assessment especially in terms of behaviour. However, the power of the matching is dependent on formalisation of the description of the new domain. Generally consider-

able detail has to be acquired before formal description can begin, hence formal specification may help the later stages of assessment but early matching may have to use informal (language based) semantics.

- Understanding – formalisation militates against understanding of modules unless the developers share knowledge of the same formal language. Given the lack of standard formal languages they may act as a barrier to understanding.
- Customisation and composition – formal specification promises to make composition easier. Customisation, however, will be discouraged as this may alter the properties of a formally specified component. The success of design by composition of rigorously design components therefore depends on the goodness of fit in the new application context.

In conclusion formal methods show some promise for tackling some of the problems of reuse, especially in the composition of new systems. However, formal methods are no panacea and, as in software development more generally, they have encountered resistance and little success. Formalisation may only be effective if it is embedded in CASE tools and the complexities of its notation and language are hidden from the software developer.

6.5.3 Domain analysis

Domain analysis attempts to address the reuse problem by an exhaustive analysis of a 'domain', i.e. a company's business or a sector of industry. The hypothesis is that by analysing all the necessary components reuse is achieved by complete anticipation. All the necessary pieces of a future system have been analysed and specified, hence reuse is a matter of retrieval and composition of the new system. Unfortunately, domain analysis is an immature field and no satisfactory methods exist. Indeed there is no definition even of what a domain is.

Reusability by domain analysis has been proposed as a solution to requirements analysis and system development by composing applications from existing, reliable components, thus delivering productivity benefits by saving development effort (Biggerstaff and Richter, 1987). Unfortunately, to date, the achievements in reuse have been modest because of a variety of management and technical problems (Prieto-Diaz, 1991):

- Existing material is not structured into easily reusable 'chunks'.
- Available information is incomplete and untrustworthy.
- Selection procedures tend to depend on exact matching of index terms which, in a large reusable component library, suggest many small unsuitable components and omit large subsystems which merely share similarities.
- Managerial and social barriers hinder the uptake of effective software reuse.

Domain analysis necessitates some mechanism for component retrieval. Most practitioners have used faceted classification schemes borrowed from information science to index specifications in the belief that sophisticated descriptors can deliver accurate and appropriate retrieval (Prieto-Diaz and Freeman, 1987; Boldyreff, 1989). Unfortunately, retrieval efficiency is limited by the lack of a universal thesaurus for describing software specifications from many domains. Furthermore, the effectiveness of reusability is dependent on software engineers documenting their specifications with agreed terms. Experience in the PRACTITIONER project (Boldyreff, 1989) has suggested that term definition is difficult and evidence for industrial take-up is not encouraging.

Whilst considerable research is currently focused on the use of domain knowledge in reuse, little thought has been given to the practical problem of initially eliciting such knowledge. Deriving application knowledge through domain analysis can be difficult and time consuming (Arango, 1987; Prieto-Diaz, 1991). Indeed (Freeman, 1987) reports that the experience of the DRACO project showed that the effort of capture and specification outweighed the potential benefit for reuse, even in a well-structured domain of oil exploration engineering.

Domain analysis has several problems which will have to be solved if it is to be effective for reuse. To compare this approach against the standard set of issues:

- Retrieval – the principal approach here is faceted classification, but this suffers from the lack of a thesaurus for software engineering and domain terms. Classification is also a time consuming task.

- Matching and assessment – domain analysis gives little or not help here beyond that afforded by keyword classification. Matching may be considered as a problem avoided if the reusable components are already present in the domain specification, although this does not avoid the problem of how well they fit a new context.

- Understanding – the complete structure of a domain analysis may help component understanding but this will depend on the overall quality of that documentation.

- Customisation and composition – reuse is usually by customisation although no direct help is given to guide the developer. Domain analysis may help composition if a large superstructure of a system is available. Again no direct support is given, so the developer has to rely on current methods and CASE tools.

Overall domain analysis does not appear to be promising. Athough is recent approach and ideas are still developing, there are significant problems with method and tool support. Currently domain analysis appears to be a 'grown-up' version of systems analysis and uses standard methods such as SSA (Yourdon, 1989). Likewise tool support is limited to current CASE technology. Perhaps a more significant and worrying problem for reuse is the idea that domain analysis can anticipate the future. All systems change in response to their environment, therefore unless a domain specification is continually updated it will become progressively out of date and hence less suitable for reuse.

6.5.4 Template approaches

Template approaches owe some of their heritage to the success of generalised application packages which provide a software template which can be parametrised according to the customer's needs. Interest in this area is growing as 'vertical market' software becomes more prevalent.

Templates are related to the domain analysis theme with the addition of the hypothesis that applications fall into generalised categories from which templates can be constructed as starting points for development (Reubenstein, 1990, Harandi and Lubars, 1987). The template approach comes in variety of flavours reflecting different levels of ambition, as illustrated in Figure 6.3. Evolutionary within-domain reuse is currently being explored in the ITHACA project in which application frames and generic objects are being developed for restricted domain classes e.g. public administration (Funghi et al., 1991). However, while templates may provide partial reuse of domain knowledge no complete or satisfactory set of domain templates has been proposed. Templates, or cliches, pose a further problem. If templates are too abstract they require considerable customization effort while too specialised templates will have a limited scope in target domains for reuse. Furthermore, generalised templates create a selection problem because choosing which template is appropriate for a new application may not be easy.

To maximise benefits software reuse needs to be effective in requirements engineering. Automated support for reuse in requirements engineering can enable reuse of previous solutions in new problem domains, thereby facilitating a known determinant of expert performance (Guindon and Curtis, 1988). Templates can be employed for specification reuse by providing generalised domain descriptions, or for design reuse by providing generic problem solutions (Shaw, 1991). Template-driven specification reuse may also give leverage on the general problem of factual ambiguity in requirements engineering, acknowledged to be one of the most error prone and ultimately costly phases of systems development. (Balzar et al., 1983). Reuse is at the abstract level of problem classes, in which the knowledge about a problem and its potential solution is transferred across domain boundaries. This approach of abstraction and provision of general solutions poses a considerable matching problem when reuse is scaled up, either in terms of numbers of templates in one domain or in terms of facilitating inter-domain reuse.

(a) Generalised template customised during development

(b) General Application Frame acts as system structure-design by composition and specialisation of generic objects

abstract template of domain class

specialise template for new application

jig-saw reuse

-system boilerplate

generic objects fitted into application frame

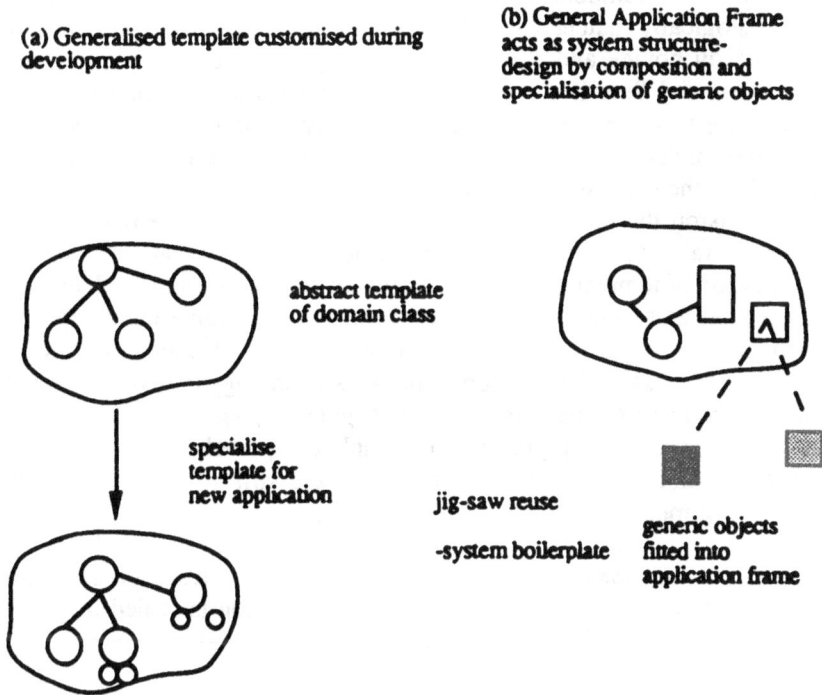

Figure 6.3 Different approaches to template reuse

One solution is to use analogical reasoning to match specifications. Several studies have demonstrated that analogical reasoning is an important determinant of analysts' expertise (Vitalari and Dickson, 1983), and analogical mapping of problem domains has been demonstrated to be computationally tractable (Holyoak and Thagard, 1989; Hall, 1989). Different types of analogy have been proposed, e.g. purpose directed (Karakostas, 1989) derivational (Carbonell, 1985) and structural (Gentner, 1983). Although no consensus has been reached, most authors agree that analogy involves a mapping of the properties of one problem space to another, similar, problem space (Carbonell, 1985).

The starting point for all analogy is recognition of salient features shared by two problem spaces. A common property of analogical definitions is the notion of mapping between structures, i.e. organised collections of primitive objects. Therefore, analogy has the potential to solve two major problems in reusability: (i) retrieval of specifications from libraries by inferring the appropriate match between properties of a target context and knowledge contained within reusable specifications and software descriptions; (ii) matching potentially reusable components to their new application context by supporting reasoning about similar properties of components and, more importantly, reasoning about models consisting of many components. Experimental studies have proved the hypothesis that analogically matched specifications can be reused leading to improved performance by novice and expert software engineers (Sutcliffe and Maiden, 1990; Maiden and Sutcliffe, 1991a,b).These studies have demonstrated that reuse will have to solve the understanding problem, implying the analogy must be thoroughly understood to prevent errors in transfer. Active explanation will be essential for the success of any reuse support tool.

In conclusion the template approach shows some promise but has to surmount considerable problems. No theory of domain abstraction exists to define the composition of templates or what a sufficient set of abstract templates should be. Limited progress may be possible within specific domains but to enable more wide ranging reuse a set of abstract templates and associated matching processes will be necessary. Application of theories of analogy may offer a way forward for the process of matching and for suggesting suitable abstractions for template libraries. Comparing the overall template approach with the framework issues the situation is less clear, as templates are still an immature research approach. No immediate panacea can be promised.

- Retrieval – this may not be a problem if the template library is small and if reuse is restricted within a domain, but as ambition is scaled up retrieval will become a significant problem.

- Matching and assessment – this is a significant problem even for small-scale use of templates. The needs of a new application context may match to all or only part of an existing template. Matching may be difficult to determine from inspection of an abstract template and tool support will be necessary. At present there is little help for this task.

- Understanding – in more concrete application templates understanding may not be a serious problem; however, for abstract templates understanding can become a serious barrier (Sutcliffe and Maiden, 1991). Explanation facilities will be necessary.

- Customisation and Composition – customisation tends to be more usual with templates; however, application frames may be.composed by placing generic objects in the boilerplate of an application. Little methodical or tool support is available although research projects are addressing this issue (Constantinopolous et al., 1990).

In conclusion, templates do appear to offer a way forward by delivering reuse through large-scale abstract structures. However success will only be delivered if the matching problem can be solved. As structures become larger the chance of exact matching decreases. Mismatches may occur because of functionality, structure or just the different scale of structures. Templates can be seen as approaching reuse by a collection of objects thereby creating some higher order composite. Matching may be easier with smaller components. Success in matching and understanding abstract templates for inter-domain reuse will require application of sophisticated AI techniques (Maiden and Sutcliffe, in press), however, more modest intra-domain reuse may be achieved in the short term.

6.6 CONCLUSIONS

The outlook for reuse is not promising. Optimistically, the considerable effort put into object oriented methods and tools may start to bear fruit, but this does not address the vast mass of existing software, unless a reverse engineering exercise is undertaken on the software maintenance mountain. Formal methods cannot provide a silver bullet for object orientation or specification of modular software.

Probably two types of reuse achievement may develop. In the short term solving the managerial, social and critical mass problems will empower reuse, especially creation of libraries of object designed for reuse. Success here will depend not only on design quality but also on overcoming the usual barriers within the software industry of incompatible languages, operating systems, etc. Reuse may spread in small well-defined sectors where libraries can be constructed and economically exploited. Indeed this is already happening in interface software and C++; for instance, window management and user interface management system software.

Spread into other domains may be hindered by the take-up of object oriented languages and methods. If the history of structured languages (e.g. COBOL85 and RPG3) and structured methods is repeated this may take some time. Reuse in this case will make design into an exercise in composition. The more disciplined and formal the designed components are, the more development can be empowered by 'off the shelf' building blocks. This philosophy appears to be more prevalent in Japan than in USA and Europe.

Another competing but probably ultimately compatible approach may come from application of knowledge based systems to CASE technology to enable reuse by templates. Small-scale success within domains appears to be possible from current R and D projects. Also considerable interest continues in vertical market software which links concepts from domain analysis and templates. More ambitious problem oriented reuse of abstraction may be possible with more sophisticated, intelligent 'reuse advisor' tools. The template approach has the added attraction that abstractions and reasoning tools can act as a bridge to match new applications to existing software specification and designs. This may unlock the vast quantity of already developed applications for potential reuse.

REFERENCES

Arango, G. 1987, Evaluation of a Reuse-based Software Construction Technology, internal document, Department of Information and Computer Science, University of California, Irvine.

Ballin, S.C. 1989, An object oriented requirements specification method. Communications of the ACM, **32**, 608-623.

Balzer, R., Cheatham, T.E. and Green. C. 1983, Software technology in the 1990s: using a new paradigm, IEEE Computer, November 1983, pp. 39-45.

Bauer, D. 1991, The Bolbingen building blocks: A catalogue of reusable software components, in Proceedings of the 1st International Workshop on Software Reuse, Dortmund, Germany, July 1991, (eds. Prieto-Diaz, R., Schaffer, W., Cramer, J. and Wolff, S. STW Memo, University of Dortmund, Germany.

Biggerstaff, T.J. and Richter, C. 1987, Reusability framework: Assessment and directions, IEEE Software, March 1987, pp. 41-49.

Boldyreff, C. 1989, Reuse, Software Concepts, Descriptive Methods and the Practitioner Project, ACM SIGSOFT *Software Engineering Notes*, **14** (2), April 1989, pp. 25-31.

Booch, G. 1986, Object oriented development, IEEE Transaction on *Software Engineering*, SE-12, (2), pp. 211-221.

Booch, G. 1991, *Object Oriented Design with Applications*, Benjamin Cummings.

Burstein, M.H. 1988, Incremental learning from multiple analogies, in *Analogica* (Research Notes in Artificial Intelligence), edited by A.E. Prieditis, Pitman, London, pp. 37- 62.

Carbonell, J.G., 1985, Derivational analogy: a theory of reconstructive problem solving and expertise acquisition. Technical report CMU-CS-85-115, Computer Science Department, Carnegie-Mellon University, Pittsburgh, March 1985.

Coad, P. and Yourdon, E. 1990, *Object Oriented Analysis*, Yourdon press, Prentice Hall, NY.

Constantopoulos, P., Jarke, M., Mylopoulos, J. and Vassiliou, Y. 1991, Software Information Base: A Server for Reuse, Technical report, FORTH Research

Institute, University of Heraklion, Crete.

Cook, S. 1986, Languages and object oriented programming, *Software Engineering Journal*, 1,(2), pp. 73-80.

De Marco, T. 1978, *Structured Analysis and System Specification*, Prentice-Hall.

Doerry, E., Fickas, S., Helm, R. and Feather, M. 1991, A model for composite system design, in Proceedings of 6th International Workshop on Software Specification and Design, Como, Italy, IEEE Press.

Falkenheimer, B., Forbus, K.D. and Gentner, D. 1987, The structure-mapping engine: algorithm and examples, TR No. UIUCDCS-R-87-1361, Department Computer Science, University of Illinois at Champaign.

Finkelstein, A., 1988, Re-use of formatted requirements specifications, *Software Engineering Journal*, September 1988, pp. 186-197.

Frakes, B. 1991, A survey of software reuse, in Proceedings of the 1st International Workshop on Software Reuse, Dortmund, Germany, July 1991, (eds) Prieto-Diaz R., Schaffer, W., Cramer, J. and Wolff, S. STW Memo, University of Dortmund, Germany.

Freeman, P. 1987, A Perspective on Reusability: software reusability, IEEE Tutorial, IEEE Press.

Funghi, M.G., Guggino, M. and Pernici, B. 1991, Reusing Requirements through a Modelling and Composition Support Tool. In Proceedings of CAiSE 91 – Advanced Information System Engineering. (eds.) Andersen, R., Bubenko, J., and Solvberg, A. *Lecture Notes in Computer Science*, No. 498, Springer-Verlag, Berlin.

Gentner, D. 1983, Structure-mapping: a theoretical framework for analogy, *Cognitive Science*, 7, pp. 155-170.

Gick, M.L. 1989, Two functions of diagrams in problem solving by analogy, Knowledge acquisition from text and pictures, (ed.) H. Mandi and J.R. Levin, Elsevier Science Publishers B.V. (North-Holland), pp 215-231.

Gick M.L. & Holyoak K.J., 1983, Schema induction and analogical transfer, Cognitive Psychology 5, pp. 1-38.

Gougen, J. 1986, Reusing and interconnecting software components, IEEE Computer, February 1986.

Guindon, R. and Curtis, B. 1988, Control of cognitive processes during software design:What tools are needed?, Proceedings of CHI 1988 conference: Human Factors in Computer Systems, edited by E. Soloway, D. Frye and S.B. Sheppard, pp. 263-269, ACM Press.

Guttag, J.V., 1975, The specification and application to programming of abstract data types, Ph.D. Thesis University of Toronto.

Hall, R.P. 1989, Computational approaches to analogical reasoning: a comparative analysis, *Artificial Intelligence*, **39**, pp. 39-120.

Harandi, M.T. and Lubars, M.D. 1987, Knowledge-based design using design schemas, Proceedings of the 9th International Conference on Software Engineering, April 1987, pp. 253-262.

Holyoak, K.J. and Thagard, P. 1989, Analogical mapping by constraint satisfaction, *Cognitive Science*, pp. 295-355.

Jackson, M.J. 1983, *Systems development*, Prentice-Hall International.

Jacobsen, I. 1987, Object oriented Development in an Industrial Environment, in Proceedings of OOPSLA-87, pp. 183-191, ACM Press.

Jacquart, R. 1991, Reuse of formal development, in Proceedings of the 1st International Workshop on Software Reuse. Dortmund, Germany, July 1991, (eds) Prieto-Diaz R., Schaffer, W., Cramer, J. and Wolff, S.. STW Memo, University of Dortmund, Germany.

Jansweiller, W., Elshout, J.J., and Wielinga, B.J. 1989, On the multiplicity of learning to solve problems, in 'Learning and Instruction. European Research in an International Context.' vol II and III, (ed.) Mandel, H., de Corte, E., Bennet, N. and Friedrich, H.F. Oxford, Pergamon.

Johnson, W.L. 1990, Understanding and debugging novice programs, *Artificial Intelligence*, **42**(1), pp. 51-97.

Karakostas, V. 1989, Requirements for CASE tools in early software reuse, ACM SIGSOFT *Software Engineering Notes*, **14**, (2), pp. 39-41.

Lewis, M.W. and Anderson, J.R. 1985, Discrimination of operator schemata in problem solvers, *Journal of Experimental Psychology: Learning, Memory and Cognition*, **8**, (5), pp. 484 - 494.

Longworth, P.G. and Nicholls, D. 1986, *SSADM – Structured Systems Analysis and Design Method*, NCC Publications.

Luqi, 1989, Knowledge-based support for rapid software prototyping, IEEE Expert, Winter 1988, 9-18.

MacDonald, I.G., 1986, Information engineering: An improved, automatable methodology for the design of data sharing systems, in Information systems design methodologies: Improving the practice, Olle, T.W., Sol, H.G., and Verrijn-Stuart, A.A. (eds), North Holland, Amsterdam.

Maiden, N.A.M. 1991, Analogy as a paradigm for specification reuse, *Software Engineering Journal* 6(1), pp. 3-15.

Maiden, N.A.M. and Sutcliffe, A.G. Analogous matching for specification reuse. CACM, **35** (4), pp. 55-64.

Maiden, N.A.M. and Sutcliffe, A.G. manuscript in preparation(a), The abuse of reuse: why software reuse must be taken into care.

Maiden, N.A.M. and Sutcliffe, A.G. manuscript in preparation(b), Cognitive models of expert software reusers.

Meyer, B. 1988, *Object Oriented Software Construction*. Prentice Hall, Englewood Cliffs, NJ.

Miyake, N. 1986, Constructive interaction and the iterative process of understanding, *Cognitive Science* **10**, pp. 151-177.

Moineau, T. and Gaudel, M.C. 1991, Software reusability through formal specifications, in Proceedings of the 1st International Workshop on Software Reuse, Dortmund, Germany, July 1991, (eds) Prieto-Diaz, R., Schaffer, W., Cramer, J., and Wolff, S. STW Memo, University of Dortmund, Germany.

Nanja, M. and Cook, R.C. 1987, An analysis of the online debugging process, in 'Empirical Studies of Programmers: Second Workshop', (ed.) Olson, G.M., Sheppard, S. and Soloway, E., Ablex, pp. 172-184.

Novick, L.R. 1988, Analogical transfer, problem similarity and expertise, *Journal of Experimental Psychology: Learning, Memory and Cognition*, **14**, 3, pp. 510-520.

Prieto-Diaz, R. 1991, Implementing faceted classification for software reuse,

Communications of the ACM **34** (5), pp. 88-97.

Prieto-Diaz, R. and Freeman, P. 1987, Classifying software for reusability, IEEE Software, January 1987, pp. 6-16.

Reubenstein, H.B. 1990, Automated Acquisition of Evolving Informal Descriptions, Ph.D. Dissertation (A.I.T.R. No. 1205), Artificial Intelligence Laboratory, Massachusetts Institute of Technology.

Robinson, P.J. (ed.) 1987, The HOOD Manual, Issue 2.1, European Space Agency, Noordwijk, Netherlands.

Rumbaugh, J., Blaha, M., Premerlani, W., Eddy, F. and Lorensen, W. 1991, *Object Oriented Modelling and Design*. Prentice Hall, Englewood Cliffs, NJ.

Shaler, S. and Mellor, S.J. 1988, *Object Oriented Systems Analysis*. Yourdon press.

Shaw, M. 1991, Hetrogeneous Design Idioms for Software Architecture, in Proceedings of 6th International Workshop on Software Specification and Design, Como, Italy, IEEE Press.

Sutcliffe, A.G. 1988, *Jackson System Development*. Prentice Hall, Englewood Cliffs, NJ.

Sutcliffe, A.G. 1991a, Object oriented system development: a survey of structured development methods. *Information and Software Technology*, **33** (6), pp. 433-442.

Sutcliffe, A.G. 1991b, Object oriented systems analysis: the abstract question. in Proceedings of IFIP working group 8.1. Conference on Object oriented approaches in Information System Development. (eds.) Van Assche, F., Moulin, B. and Rolland C. pp. 23-37, North Holland.

Sutcliffe, A.G. and Maiden, N.A.M. 1990a, Specification reusability: why tutorial support is necessary, in Proceedings of SE-90, Brighton, UK, July 1990.

Sutcliffe, A.G. and Maiden, N. 1990b, Software reusability: delivering productivity gains or short cuts, in Proceedings of Interact-90, (eds.) Diaper, D., Gilmore, D., Cockton, G. and Shackel, B. pp. 895-902, North Holland.

Sutcliffe, A.G. and Maiden, N., in press, Analysing the Analyst: cognitive models in software engineering, to appear in *International Journal of Man Machine Studies*.

Sutcliffe, A.G. and Maiden, N.A.M., in press, Analysing the analyst: cognitive models of software engineers, submitted for publication.

Tracz, W. 1991, The Impact of Domain Analysis on Software Reuse, in Proceedings of the 1st International Workshop on Software Reuse, Dortmund, Germany, July 1991, (eds.) Prieto-Diaz, R., Schaffer, W., Cramer, J., and Wolff, S. STW Memo, University of Dortmund,Germany.

Uhl, J. and Schmid, H.A. 1990, A systematic catalogue of reusable abstract data types. *Lecture Notes in Computer Science*, No. 460, Springer-Verlag, Berlin.

Van Assche, F., Layzell, P., Loucopoulos, P. and Spelincx G. 1988, Information systems development: a rule-based approach, in Proceedings of International Workshop on Knowledge-based Systems in Software Engineering, UMIST, UK, March 1988.

Vitalari, N.P and Dickson, G.W. 1983, Problem solving for effective systems analysis: an experimental exploration. CACM, **26** (11), pp. 948-956.

Walton, P. 1991, Software reuse: management issues, in Proceedings of the 1st International Workshop on Software Reuse, Dortmund, Germany, July 1991, (eds.) Prieto-Diaz, R., Schaffer, W., Cramer, J., and Wolff, S. STW Memo, University of Dortmund, Germany.

Wasserman, A., Pircher, P.A. and Muller, R.J. 1989, Concepts of object oriented design, Interactive Development Environments Report, San Francisco, CA.

Wirfs Brock, R.J., Wilkerson, B. and Wiener, L. 1991, *Designing Object Oriented Software*, Prentice Hall, Englewood Cliffs, NJ.

Wirsing, M. and Hennicker, R. 1991, Formal aspects of reusability, in Proceedings of the 1st International Workshop on Software Reuse, Dortmund, Germany, July 1991, (eds.) Prieto-Diaz R., Schaffer, W., Cramer, J., and Wolff, S. STW Mémo, University of Dortmund,Germany.

Young, R.M. 1983, Surrogates and Mappings: two kinds of conceptual mappings for interactive devices, in *Mental Models*, (eds.) Gentner, D. and Stevens, A.L. Lawrence Erlbaum Associates, pp. 35-52.

Yourdon, E. and Constantine, L. 1977, *Structured design*. Yourdon Press, New York.

Yourdon, E. 1990, *Modern Systems Analysis*, Prentice Hall.

7 Formal Methods and Transformations in Software Reuse

Martin Ward
Centre for Software Maintenance

7.1 INTRODUCTION

Production of software is costly and error prone, and the most important means of production (good programmers) are scarce. Therefore there exists a need to circumvent this costly manual production process. Analogies from classical engineering suggest that by building up a catalogue of standard components and construction techniques, whose characteristics are well documented, the cost of new construction projects can be greatly reduced. The bridge builder knows under what conditions a 'suspension bridge' is the right approach and has a collection of standard girders, cables, nuts, bolts etc. which s/he can use in the design. This has lead to the notion of a component repository as a means to reduce the effort involved in constructing new software.

7.1.1 Current reuse technology

The desire to avoid writing the same section of code more than once led to the invention of macros and subroutines. These allow the reuse of common code sequences, but the reuse is confined to a single author, or at most a single project. This is too restricted to bring relief to the industry.

Standard subroutine libraries have proved a more powerful technique. Packages like SSP and SPSS have had high success because they not only relieve the programmer from the drudgery of coding but also (in their limited domain of application) relieve him from the need to develop an algorithm, or to understand in detail the theory behind it. Unfortunately, only limited progress has been made in this area since the early days.

We could conceive of a high-level language as an attempt at reusability: canonical structures which frequently occur in a program, such as looping constructs and methods of procedure call, have been encapsulated into a single command. In addition, common programming *techniques* such as register allocation, loop strength reduction and other optimisations are carried out automatically by the compiling system. In the case of the GNU C compiler [8] 'function inlining' can be carried out automatically: the *compiler* selects suitable procedures for inline expansion. This means that the distinction between macros and procedures has (for most practical cases) been removed. The further extension of this idea, to still higher-level languages, is restricted by the perceived need for compilation to be a totally automatic process.

The module or package concept in languages such as Modula-2 or ADA appears to provide even greater support for reuse. The programmer can define data or procedural abstractions and link code of a reasonably general nature into the code she is writing. However, the module implementations at hand are often incompatible with each other since they have been developed independently. They will also be incompatible with the new product under construction. The difficulties involved in re-writing and patching existing modules, without introducing bugs, can be greater than the cost of starting from scratch.

Certain operating system features, such as pipes, have been considered as a means of supporting reuse [5]. A new system is built up by combining programs, taken from a 'toolkit' of standard utilities, using these features. This can be very effective for developing simple file processing utilities; but for more complex programs, especially those which do not fit the simple unidirectional dataflow model provided by pipes, the method is less suitable. Also, constructing a program out of components which are complete programs or processes can result in unacceptably poor performance.

In general, the 'component library' approach to reuse presupposes the availability of a large catalogue of high-quality, general-purpose modules which can be made available to developers of new products. A problem with such a library is that for it to be used effectively the programmer has to know what is available and what each piece of code does, and must be able to combine them on the source code level without any further support from the system. This becomes extremely difficult as the component library gets larger, but that is just when it is becoming potentially more useful. As a result these methods have found their greatest success when they are limited to a narrow domain of programs. A 4GL[1] can be seen as an example of a collection of reusable modules for a narrow programming domain, together with the means to compose them into new programs.

A further problem with component libraries containing code modules is that only a minor part of the program development effort is spent on coding, therefore we want to reuse more than just the code. Specifications, development methods, designs and documentation should all be reusable.

7.1.2 Traditional development methods

The traditional development methods can be grouped into four main types:

- The traditional 'waterfall' life cycle which starts with a fixed specification and develops it into a finished product through a number of stages.

---

1. 4th Generation Language

- Incremental development, in which a small part of the product is implemented and then enhanced as the specification is developed.
- Rapid prototyping, in which a prototype of the main part of the system is developed and analysed and used as the basis for the next in a series of prototypes. The traditional waterfall method has been described as 'slow prototyping'.
- A combination of the above.

All of these methods can be seen as applications of reuse: the initial work (a specification, a partial implementation or a prototype) is reused in the later stages. Software maintenance can be seen as the development of enhanced products involving a high degree of reuse of the existing product. However the reuse is almost invariably restricted to a single project, and is often restricted to code reuse.

We want to extend reuse to cross project boundaries and to extend the base of components which can be reused to include all the products of development work.

7.2 THE SOFTWARE REPOSITORY

The idea of constructing new software by composition from a collection of reusable components is not new and clearly has many attractions. However, it has yet to receive widespread implementation. There appear to be several technical reasons for this (in addition to the managerial issues such as the 'Not Invented Here' syndrome):

- The repository must be large enough to contain a useful collection of components, yet each component must be readily accessible.
- The components must be highly reliable since they will (hopefully) be reused in many applications.
- There must be some means of extracting components from existing code for addition to the archive: writing a complete library of components from scratch would involve a great deal of investment of effort before any return on the investment would be perceived.
- In order to be widely useful the components should be written to handle the most general cases; this means that programs constructed from components can be much less efficient than programs written from scratch which can exploit regularities in the data.

This paper describes how the theory of program refinements and transformations developed in [10, 12, 16, 18] can be applied to the construction of a repository of usable components from which new software can be constructed. The repository contains code, specifications and techniques as the components, connected by formal and informal links. The formal links record proven knowledge about the components, for example an abstract specification will be connected via a *refinement link* to its implementation, two algorithms for solving the same problem will be connected via a *transformation link* and an implementation of an abstract data type in terms of concrete data types will be recorded as a *reification link*. Informal links will enable keyword searches and will connect informal text descriptions of components to other components.

7.3 THEORETICAL FOUNDATION

In this section we will briefly outline the theoretical work behind the wide spectrum language and its transformation theory.

7.3.1 The Wide Spectrum Language WSL

A Wide Spectrum Language is a language which covers the whole spectrum of operations from low-level programming constructs (loops, assignments, **goto's** etc.) to high level abstract specifications expressed in non-executable form. In [10, 12, 18] a wide spectrum language (called WSL) is developed specifically to form the foundation for a transformation theory. A refinement of a program in WSL is another program which is more defined (i.e. is defined on a larger initial set of states) and more deterministic (i.e. for each initial state it has a smaller set of potential final states). In practice, this means that any component of a program can be replaced by any refinement of that component without affecting the operation of the program as a whole. A transformation is a refinement which is correct in both directions: applying a transformation to a component of a program will change it in such a way as to preserve the semantics[2] of the program.

The combination of refinements and transformations with a wide spectrum language means that the *implementation* of a specification is just a special case of refinement. We don't have to worry about separate specification and programming languages and the translation between them: within this (formal) method, all operations are carried out in a single (formal) language.

In [13, 16] we have developed this theory into a formal method for extracting specifications from program source code, which we can prove to be correct representations of the source code. In the other direction, we can refine specifications into (provably correct) implementations: see for example [14, 15, 17]. In the ReForm project (a joint project between the University of Durham, CSM Ltd. and IBM UK Ltd., partially funded by the DTI, [2,2,19]) a prototype program transformation tool is being developed which has demonstrated that these methods are practical for large commercial software systems.

---

2. The semantics of a program is a mathematical function which captures the external behaviour of the program while ignoring its internal details.

The research project described in this paper aims to apply these methods to the construction of component repositories. The aim is to extent the component-reuse metaphor to higher levels of abstraction (just as eForm aims to extend maintenance to higher levels of abstraction). The formal method is essential for software to be reused successfully: we should only need to look at the high-level specifications of a component to decide whether it meets our requirements, and for this decision to be made correctly we require a high degree of confidence in the correctness of the implementation.

7.3.2 The atomic specification

The atomic specification provides a notation for specifying the input – output relation of a program using logical formulae while abstracting away all implementation details:

Definition 1 *The Specification Statement: written x/y.Q, where Q is a formula of first order logic and x and y are sequences of variables, is a form of non-deterministic assignment statement. Its effect is to add the variables in x to the state space, assign new values to them such that Q is satisfied, remove the variables in y from the state and terminate. If there is no assignment to the variables in x which satisfies Q then the atomic specification does not terminate (i.e.) it is not defined for those initial states).*

This is based on the 'atomic description of Back [1].

The atomic specification is *non-deterministic* in its effect: if there is more than one assignment of values to the x variable which satisfy Q then we do not specify any particular choice or choice method. A program which has less non-determinism is a perfectly valid refinement of the specification. Non-determinism in specifications is important in avoiding implementation bias: it provides a mechanism for recording the set of acceptable results, and the freedom available to the implementor.

Some examples of atomic specifications:

1. <x>/<>.(x>y)

This sets x to any value greater then the value of y. If there is no such value then the specification does not terminate.

2. <z>/<>.(z=x+y); <>/<z>.(x=z)

This sequence implements the assignment statement x:=x+y using temporary variable z which is not used elsewhere.

3. $\approx '/<>.Q; \dot{x}/\dot{x}'.(\dot{x}=\dot{x}')$

Here x is a sequence of variables and x' a sequence of temporary variables. These statements implement the general assignment statement: x:=x'.Q which assigns new values x' to x where Q gives the relation between x and x'.

4. $<x>/<>.(x\_\leq y<(x+1)\_)$

This specifies an 'integer square root' function without giving the algorithm to achieve it.

7.3.3 The *join* construct

Together with the atomic description and more familiar programming constructs the Wide Spectrum Language includes a new construct called *join*. This provides a mechanism for composing specifications out of less-specific components. The join of two programs is defined to be the weakest (i.e. least defined) program which satisfies any specification satisfied by either of the two programs. If one of the component programs does not terminate for a particular initial state then it cannot satisfy any specification defined on that state, so the join of the two programs is identical to the other program on that state. A property of the join construct is that any program which refines both components will also refine their join. This property is very useful is searching the repository: if we have a specification which we wish to implement we first want to search the database for an equivalent (or at least similar) specification which has already been implemented. For a large and complex specification this will give rise to a potentially highly complex matching problem. If, however, the specification is expressed as the join of several simpler specifications then the matching problem for each component will be much easier to solve. Once we have found all the components, we can search through the refinement links to find a common ancestor to all of the components. From the above property of join, this ancestor will be a correct implementation of the full specification. We give an example of this search below:

In general, the components to be joined will each have a high degree of non-determinism, the process of joining them together restricts the hon-determinism to be consistent with both: for example the join of $<x>/<>.(x\_ \leq y)$ and $<x>/<>.((x + 1)\_ > y)$ is the integer square root function discussed above. Joining two inconsistent programs together may result in an unimplementable specification: for example, the join of x := 1 and x := 2 is a program which assigns x the value 1 and also assigns x the value 2. This is termed a *null* program (called a 'miracle' by some authors [7] after Dijkstra's 'Law of Excluded Miracle' [3]) which is technically a correct refinement of any specification, but which (unfortunately) cannot be implemented on a machine! They are however an important theoretical tool.

We have used the wide spectrum language and its transformation theory to develop tools for the development of algorithms and programs from specifications, and the derivation of specifications from code (we term this process 'inverse engineering'). A large catalogue of practical transformations and refinements has been developed which are being applied to a wide variety of programs.

7.4 WHY INVENT ANOTHER NEW LANGUAGE?

There are several reasons why we have invented another language rather than use an existing programming language such as C or ADA:

- We needed a language with simple semantics and tractable reasoning methods. In particular, our language has been designed from the start with ease of transformation and refinement as a major objective. New constructs are added to the language only if we can show that they will be easy to work with. We need a useful set of transformations which make use of that construct before it becomes part of the language. This policy has proved very successful and enabled us to avoid some of the problems which can occur when the language definition is the starting point for research.
- We wanted to include the implementation of a (possibly non-executable) specification as an allowable refinement step. We also wanted to be able to write programs using a mixture of specifications and programming constructs. This facilitates the stepwise refinement of specifications into programs and the iterative analysis of programs into specifications. No existing implementable programming language includes general specifications in its syntax (for obvious reasons!).
- By expressing our results in a general language we get results which are independent of any particular programming language. Programs in existing programming language can be transcribed into WSL, manipulated as WSL programs, and then re-transcribed, perhaps into a different programming language.
- All existing programming languages have limitations (in particular, the limitation to executable constructs which is intolerable in a specific language). Also many popular languages have a number of quirks and foibles which would greatly complicate the semantics while adding little expressive power.

7.5 THE REPOSITORY

The repository consists of components connected by formal and informal links.

7.5.1 Components

The components in our repository include the following:

- Modules of code in various programming languages.
- Modules of code-level WSL: these are in a form which is suitable for automatic translation into various high-level languages or assembler.
- Modules of specification-level WSL: these are at various levels of abstraction, from Z [6] or VDM [4] style specifications to detailed algorithms.
- Derivation histories: these record the sequence of transformations and refinements which connect a specification to its implementation. Transformations and refinements are recorded in an extension to WSL called *meta*-WSL. These derivation histories can be generalised into derivation *strategies* which are meta-WSL programs.
- Meta-WSL programs: these record strategies and development techniques, including generalisations of actual development histories. Since meta-WSL is an extension of WSL, meta-WSL programs can be transformed by the application of other meta-WSL programs.
- Documentation and informal requirements.

7.5.2 Component links

These are of two major types:

Formal links which record proven facts about the components and which are therefore transitive.
 These are of four different types:

1. Change in data representation: ⟶⊦▶ This links two programs which are equivalent in effect but which use different representations of the data.

2. Refinement: ⟶▶ This links a program or specification to a refinement of it.

3. Transformation: ⬌ This links two programs which are equivalent, but which may use different internal data or algorithms.
4. Reifications: ≡▶ This is similar to transformation in that it links equivalent programs, but the program to the right uses a less abstract internal data representation and less abstract algorithms and is therefore closer to an implementation.

Informal links: ⋯▶ These connect documentation, informal requirements and keywords to the other components.

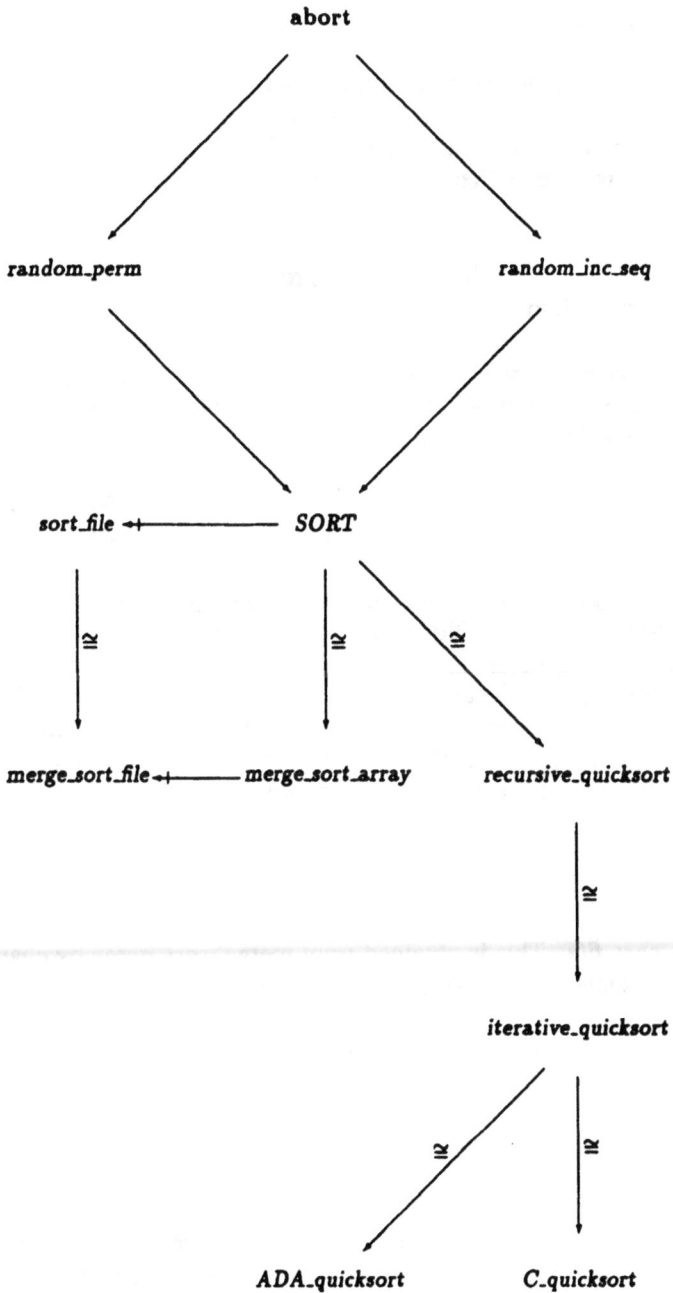

Figure 7.1 A fragment of a repository

7.6 AN EXAMPLE

In this section we present a small example of a fragment of a repository concerning sorting programs. For the moment we will restrict attention to a subset of the repository components and the formal links which connect them. The components are as follows (here A [1...*n*] is an array of elements to be sorted in place):

- abort: This is the totally undefined program; any program is therefore a refinement of abort.

- *random_perm* $=A := A'.(\exists \pi \in$ *Perms* $(n). \forall 1 \le i \le nA$ $[\pi(i)] = A$ ' [*i*]): here *Perms* (n) is the set of all permutations of the elements $\{1,...,n\}$. This program arbitrarily permutes the elements of array A.

- *random_inc_*seq $= \langle A \rangle / \langle \rangle. (\forall i, j. 1 \le i < j \le n \Rightarrow A$ [*i*] $\le A$ [*j*]). This program assigns arbitrary values to A to produce a sorted array

- SORT = **join** *random_inc_seq random_perm* **nioi**. This is a specification of a program to sort A. Note that it gives no indication of the algorithm to use (testing all possible assignments of increasing sequences to see which are permutations of the original array is not a practical sorting algorithm!). Note the use of join to split up the specification into two simpler sub-specifications. (Sort is probably a simple enough concept by itself -- this is just an example to illustrate the technique).

- *merge_sort_array*. This is the implementation of a merge sorting algorithm.

- *merge_sort_file*. This is obtained from merge_*sort_array* by changing the data representation: we use an array to represent a file.

- *recursive_quicksort*. This is obtained from SORT by 'algorithm derivation' (see below).

- *iterative_quicksort*. A reified algorithm obtained from *recursive_quicksort*.

- *ADA_quicksort*. Obtained from *iterative_quicksort* by transforming into a form which can be automatically translated into an efficient ADA module.

- *C_quicksort*. See *ADA_quicksort*.

The links which connect these components are shown diagrammatically in Figure 7.1. Let us suppose that a user of the repository wishes to implement a sorting algorithm. S/he writes his specification in the form of a join of smaller specifications and then searches the repository to see if suitable implementations already exist. The initial stages of the search could make use of informal links: for instance an analysis of the specifications would suggest that 'Array' would be a suitable keyword to restrict the search to specifications of array manipulation programs. Theorem-proving techniques can be applied to prove, for instance, that *random_perm* is a refinement of one component of the specification, and that *random_inc_seq* is a refinement of the other component. Once refinements of all components have been found then the refinement and reification links can be searched automatically to find a common descendant of all the refinements. In this case *SORT* will be found immediately.

Note that the process of finding a common descendant could fail: for example the system might notice that **skip** is a refinement of *random_perm*, but the join of **skip** and *random_inc_seq* is not fully defined. If a null program is reached in the search for a common descendant then the search has failed since such a specification cannot be implemented. This may happen because the repository is incomplete (an implementable join of the components exists but is not present in the repository), or because the decomposition of the requirements was carried out incorrectly and a null program was (inadvertently) specified.

Once *SORT* has been found, the various implementations of the specification can be extracted by following the reification links. For example the 'quicksort' implementation could be selected. This has previously been transformed into an efficient iterative algorithm which has been massaged into two forms, one appropriate for translation to C and the other for ADA. See [11] for the derivation of the programs from the *SORT* specification. Alternatively a file sorting algorithm could be extracted by following the 'change data representation' links.

This process is analogous to the process we go through when selecting a purchase from the set of manufacturers' offerings. We have a rough idea of what we want, which is still precise enough for us to check it against the given supply. Our requirements are frequently expressed as a set of objectives which we require to be simultaneously achieved: for example we may be looking for a freezer with a certain minimum capacity, temperature rating, low running costs etc. A suitable 'implementation' of these requirements has to meet them all simultaneously. As we narrow down the set of possibilities we add more details to our specification and make more precise discriminations. If no ready-built product is suitable then we may choose to get one specially manufactured. The manufacturing process will throw up requirements for components which will have to be searched for in turn.

7.7 ADDING EXISTING CODE

A major problem with the current research on reuse is that several people have produced prototype component repositories but nobody wants to start using them because of the enormous effort involved in developing a large enough set of components to populate the repository and make it usable. With the system presented here this problem is much less acute: existing code can be placed in the repository, initially with informal links only. Later, as the code is analysed using code analysis and specification tools such as the Maintainer's Assistant [9,19] smaller and more general components and their specifications will be added, together with a growing network of formal links. This process can be carried out in conjunction with normal maintenance: as the specifications of code modules are extracted they can be placed in the repository. In addition the transformational development of new code from specifications and components will provide new components and links for the repository.

7.8 PROBLEMS AND BENEFITS

7.8.1 Problems

Specification Matching: Any repository or component library is only as good as the technique for matching specifications and extracting components. This is a difficult problem in general: the problem gets more difficult as the library gets larger, but this is just when it is becoming more useful. We believe that the technique of including a large collection of 'partial' or 'generic' specifications which can be composed using the join operator will greatly assist in finding the required component and eliminating unsuitable matches. With a large set of generics there will be a number of paths through the repository, from the results of an initial informal keyword browse to the desired component. Thus the large size of the library actually assists in the search rather than hindering it. Developing a 'standard style' for writing specifications, and a standard set of generics for composing larger specifications, will greatly assist the theorem proving specification matcher and improve the ease with which specifications and other components can be extracted.

Size of Repository: The more components and links (especially formal links) there are in the repository, the more useful it will be in the construction of new software. Many of the components will be substantial pieces of code or documentation, including perhaps many different versions of the same piece of code tailored for different purposes. Thus the overall size of the repository is likely to be very large. Fortunately, the cost of magnetic storage systems is falling each year, and new technology such as optical WORM[3] devices is

---

3. Write Once Read Many

appropriate in this case because most of the operations will consist of reading from, and adding to, the repository with only very occasional deletions.

Efficiency: Constructing a program from a set of general purpose reusable components can often generate a highly inefficient result. There may be extra layers of procedure calls and the general-purpose modules are not able to make use of regularities in the data for the constructed program to carry out its actions more efficiently. This problem is minimised by the application of efficiency improving transformations to the generated code. Optimising compilers carry this out at a very low level: they construct the program from standard code blocks which implement the high-level constructs and then optimise the result to try and remove the inefficiencies introduced. The transformations we have developed work at a higher level than any optimising compiler: they include removing unnecessary procedure calls, migrating code between modules, adding data structures to store intermediate results rather than re-calculating, changing the representation of data structures, etc. Because the transformations have been proven to preserve the effect of the program, and because they can be applied and the correctness conditions checked automatically, there is no chance of introducing clerical or logical errors in a long series of transformations. Hence they can be used freely wherever appropriate to improve the efficiency of the final product to a sufficient degree. The resulting modules can in turn be added to the repository and reused, as can any new efficiency improving techniques which are developed.

7.8.2 Benefits

- Recording formal as well as informal links in the repository means that the work involved in proving that a module correctly implements its specification is not lost, but is repaid many times over.

- The repository records specifications and development methods as well as code, so these can be reused in the same way.

- Maintenance work carried out using tools such as the 'Maintainer's Assistant' [19] generates new components with validated high-level specifications as a by-product. These can be incorporated in the repository so that the existing development and maintenance investment can be made greater use of.

- The creation of formal links means that there is a high degree of confidence that the components in the repository meet their specifications; hence new programs constructed from these components will be correspondingly reliable.

- The efficiency improving transformations make it possible to construct efficient programs out of general purpose components.

- The derivation of specifications from old code undergoing maintenance means that such code can be brought into a CASE strategy.

7.9 CONCLUSION

We have described a theory for program transformation and refinement which has proved very powerful for the derivation of programs from specifications and the analysis of existing programs in software maintenance [19]. This, together with the join concept for composing specifications and programs, and the 'meta-programming' language for describing program developments, forms the foundations for the construction of a repository of reusable components which can be used in the development of new software with greater reliability at greatly reduced cost.

BIBLIOGRAPHY

[1] Back, R.J.R. 1980, Correctness Preserving Program Refinements, Mathematical Centre Tracts#131, Mathematisch Centrum.

[2] Bull, T. 1990, An Introduction to the WSL Program Transformer, Conference on Software Maintenance, San Diego, November 26th- 29th.

[3] Dijkstra, E.W. 1976, *A Discipline of Programming*, Prentice-Hall, Englewood Cliffs, NJ.

[4] Jones, C.B. 1986, *Systematic Software Development using VDM*, Prentice-Hall, Englewood Cliffs, NJ .

[5] Kernigham, B.W. 1984, 'The UNIX system and software reusability', IEEE Trans. *Software Engineering*. SE-10, September, 513- 528.

[6] McMorran, M.A. and Nicholls, J.E. 1989, 'Z User Manual,' IBM UK Laboratories Ltd., TR12.274, Hursley Park, July.

[7] Morgan, C. 1990, *Programming from Specifications*, Prentice-hall, Englewood Cliffs, NJ.

[8] Stallman, R.M. 1989, Using and Porting GNU CC, Free Software Foundation, Inc., September.

[9] Ward, M. 1988, Transforming a Program into a Specification, Durham University, Technical Report 88/1.

[10] Ward, M. 1989, Proving Program Refinements and Transformations, Oxford University, D.Phil. Thesis.

[11] Ward, M. 1990, Derivation of a Sorting Algorithm, forthcoming.

[12] Ward, M. 1990 Specifications and Programs in a Wide Spectrum Language, Durham University, Technical Report.

[13] Ward, M. 1992, A Recursion Removal Theorem, submitted to BCS Refinement Workshop, 8-11 January.

[14] Ward, M. 1990, The Largest True Square Problem – An Exercise in the Derivation of an Algorithm, Durham University, Technical Report, April.

[15] Ward, M. 1990, The Schorr-Waite Graph Marking Algorithm – An Exercise in the Derivation of an Algorithm, Durham University, Technical Report, April.

[16] Ward, M. 1991, Abstracting a Specification from Code, submitted to *Journal of Software Maintenance and Management*, August.

[17] Ward, M. 1980, Iterative Procedures for Computing Ackermann's Function, Durham University, Technical Report 89-3, February.

[18] Ward, M. 1989, A Model for Partial Programs,' Submitted to J. Assoc. Comput. Mach., November.

[19] Ward, M., Calliss, F.W. and Munro, M. 1989, The Maintainer's Assistant, Conference on Software Maintenance, Miami, Florida (October 16th-19th).

8 Domain Analysis

Pat Hall
Department of Computing, Open University

8.1 INTRODUCTION

As the breadth of application of software has been extended and we have sought more effective means of developing software, domain analysis has arisen in two different parts of software engineering:

- software reuse
- knowledge based systems.

Computing began as the mechanisation of calculation and moved from there during the 60s into the mechanisation of clerical processes. However it was realised that this was not enough, that the problem to be computerised needed to be understood within its own terms, and solutions possibly recast, if computerisation was to be successful. Thus in the early 70s we saw a concern with problem related issues, starting with structured programming and then structured analysis, leading into the rise of relational databases and the whole debate about conceptual schema that was so strong in Europe (Nijssen, 1976). This in turn led to the rise of fourth generation languages and relational databases becoming commercially available towards the end of the 70s and early 80s. Understanding a problem was seen as modelling it within the conceptual schema, or perhaps even building a simulation of the problem domain, as in the approach of Jackson (1983).

The 80s saw the ascendancy of technology, with far more powerful computing platforms becoming available. Higher quality systems were sought through the application of formal methods and mathematical proof, high resolution graphics brought CASE tools, and there was an unprecedentedly large expenditure of public money in pursuit of these software engineering goals. These technological advances also enabled the rise of expert systems and knowledge engineering approaches within narrow domains, mostly diagnostic problems rather than the more challenging configuration problems (Neale, 1988). There was an interest in knowledge elicitation and some researchers realised that there were deeper issues behind this, with the need to examine the nature of expertise and of knowledge itself (Keravnou and Johnson, 1986). Some of these methods looked outside the conventional sources of computing, into the methods of the human sciences (Suchman, 1987 and Leith, 1986, 1990 both show the intrinsic social dimension to computing technology). However, knowledge based systems have remained largely separate from conventional data processing.

In the more conventional data processing applications the breadth of applications grew and grew, with continuing problems associated with inappropriate systems failing to meet the users real needs. The traditional requirements analysis and feasibility study phases were found to be inadequate. A number of solutions arose, with participative development (Mumford and Henshall, 1979; Mumford, Land and Hawgood, 1987) having a brief fashion in the early 80s, and throughout the 80s prototyping (Boar, 1984) enabled the early delivery of a system from which to evolve a system meeting the user needs. The users were emerging as key to the success of administrative systems. System developers talked of the need to include application experts in their teams, and professional societies talked of the 'hybrid manager', both as a human bridge between human problem and system solution. IT users began to undertake IT strategy reviews, to document the business needs and from that derive the IT solutions – this usually happened as part of a consultancy exercise undertaken within the context of a pragmatic methodology, such as Information Engineering.

On the software engineering and technology side, software reuse (Freeman, 1987; Tracz, 1988; Dusink and Hall, 1990) arose as a means of improving productivity and reliability, and of developing product lines. Many companies found that they were active in the same market over many years and looked to find advantage in this. A key to success with reuse has been the recognition of the need to understand the application area within which the components will be (re)used, this activity being referred to as Domain Analysis (Neighbors, 1981;, Prieto-Diaz and Arango, 1990) – this has also been seen as important in reverse engineering (Hall, 1991). Domain analysis is also seen as a means of capturing expertise – in building the model of a domain for a product line, key expertise will have been captured, enabling product evolution more readily thereafter, untroubled by any critical dependency upon particular expert engineers and expediting the introduction of new engineers.

In this survey I will take the software reuse perspective of Domain Analysis, as typified by the collection of papers edited by Prieto-Díaz and Arango (1991). However I will also look in other areas, especially knowledge engineering, for other approaches and future directions. Figure 8.1 shows a general view of how domain analysis would related to other software development activities.

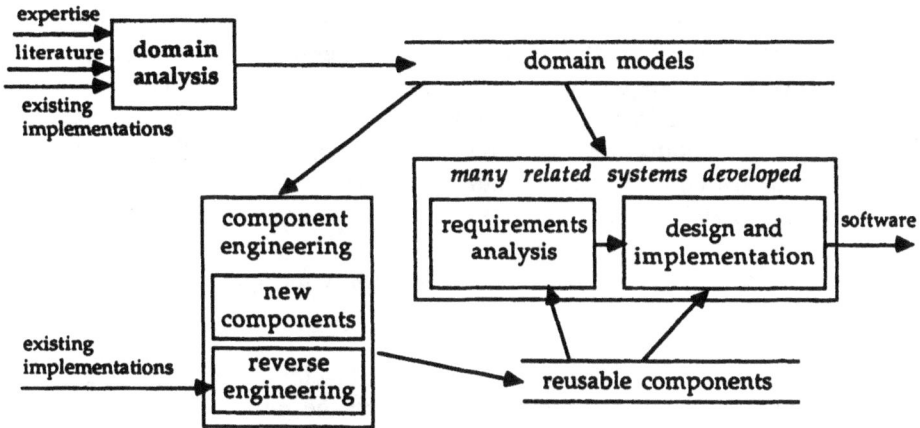

Figure 8.1 Place of domain analysis with systems development

8.2 WHAT IS DOMAIN ANALYSIS?

Domain Analysis entered software reuse through the Draco project (Neighbors, 1981), where it was defined as:

> The activity of identifying objects and operations of a class of similar systems in a particular problem domain. ... needs and requirements for a collection of systems which seem 'similar'. (p.564)

Thus domain analysis is like requirements analysis, only broader, covering a number of related systems. It is worth looking further into this paper, where we find that our task must be:

> to produce domain languages which may be transformed for optimisation purposes and implemented by software components, each of which contains multiple refinements which make implementation decisions by restating the problem in other domain languages. (p.565)

The representation of the domain becomes important, captured here as a domain language.

Since then a number of domains have been analysed – the most celebrated of these is probably the CAMP system where a number of missiles were analysed looking for common components between their controlling software, with the task of building another missile software system from these parts. It was relatively successful as a reuse project, but can also be viewed as a domain analysis project where the domain model was the set of components identified.

Both Draco and CAMP focused on implementation, though Draco (see above) recognised a sequence of languages leading up to the implementation. One strand of current development continuing this implementation orientation is domain specific architectures (DARPA, 1990), where the overall structure of a system is identified and then specialised to particular situations through the choice of particular components to slot into the overall architecture (the SEI illustrate this with a weapon launching system).

More recent approaches have then moved away form components to 'information', as in:

> ... a process by which information used in developing software systems is identified, captured, and organised with the purpose of making it reusable when creating new systems. (Prieto-Diaz, 1990, p.47)

while Arango sees this as too restrictive through the focus on systems:

> ... identification, acquisition, and evolution of reusable information on a problem domain ... (Arango, 1989)

We see then that domain analysis has moved from its component collection focus to the broad consideration of the knowledge in an area of application, in which a number of interrelated systems are deployed. Thus it can be viewed as a vehicle for documenting the expertise of engineers in the area, making the company less susceptible to loss of critical skills, as well as the more common emphasis within reuse of reducing costs and improving competitive position. Many of the case studies within this conference can be construed in this way.

8.3 DOMAIN REPRESENTATIONS

To be able to use domain knowledge effectively, we must be able to represent the results of the domain analysis so that it can be shared. In the last section we saw a quote from Neighbors in which he talks of domain languages. He did not have in mind natural language, but rather a formal language for representing the knowledge. He also envisaged:

> ... many languages at once where the languages are not in strict hierarchy. (p.566)

This is illustrated in Figure 8.2. This example, adapted from Neighbors, shows a statistical package described (designed) in terms of statistical calculation and report formatting domains. Both these domains involve some numerical (algebraic) calculation, and these three intermediate domains are defined in terms of the operating system and implementation language domains. The hierarchical decomposition is referred to as 'refinement', while each domain includes not only its domain language, but also its transformation into lower level domain languages.

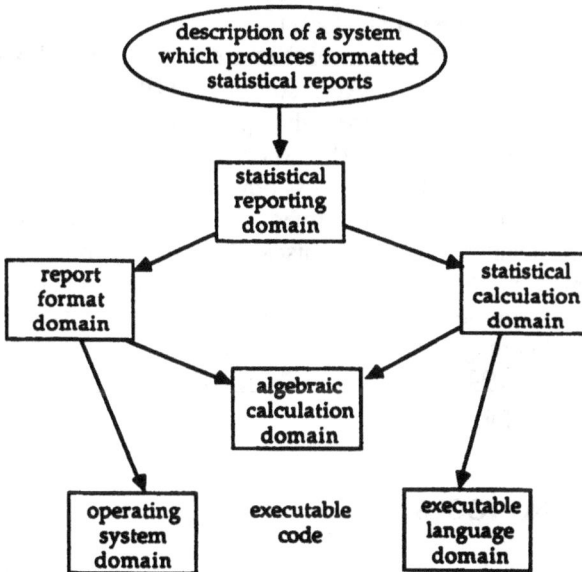

Figure 8.2 Example of a DRACO domain hierarchy
(Adapted from Neighbors, 1981)

What form could these languages take?

Clearly at the lowest level it would be program code but, as indicated above, this is too detailed. We want to represent knowledge, and for this we have many candidate representations. Damiani and Botterelli (1989) have given some requirements:

- description equivalence through terms or descriptions, using inference
- consistency of term interdependencies, checkable automatically
- ability to organise descriptions within a taxonomic hierarchy of subsumption
- ability to carry out general inferences within the model
- model size unconstrained
- ability to query the model.

The simplest candidate model formalism would be entity-relationship diagrams. Carasik et al. (1990) have objected to these on the grounds that they do not make good models of natural language. I find this argument irrelevant, but do agree that they are inadequate on the grounds that they do support hierarchy and subsumption, as required by Damiani and Botterelli above. Damiani and Botterelli use an extended entity relationship model (sometimes called a structurally object-oriented data model), in which hierarchies and inheritance have been added to entities and relationships. Underlying Damiani and Botterelli's system is a term system with rich inferencing rules. A similar system has been used in the Lassie system by Devanbu et al. (1991).

Clearly fully object-oriented approaches are candidates, following the classical paradigms such as those of Atkinson et al. (1990), Booch (1987), Cox (1986), Meyer (1988) or any of the other paradigms covered in papers like Nierstrasz (1989), Setien et al. (1989), and others in the collection of papers by Kim and Lochovsky (1989). But other less obvious approaches are also possible.

For example, on the Practitioner project (Hall et al., 1990) the thesauri used for information retrieval purposes organises the reusable concepts and in effect constitutes a domain model (see Aitchison and Gilchrist (1971) or Townley and Gee (1980) for information on thesauri). From thesauri it is then not a large jump into natural language and knowledge representation formalisms, such as Allen (1987), Horn (1991), Maarek (1990), Rau (1987) or the many others shown in the bibliography.

8.4 METHODS AND TOOLS

We have seen the objectives of domain analysis in section 2, and representations for domain models that we could use in section 3. But how do we create domain models? Current approaches are *ad hoc,* depending upon expertise and experience. Could we find a methodology for domain analysis which people new to the area could use? But:

> ... domain analysis is very hard. Typically it takes an expert in a particular problem area four months to complete a first attempt at the domain ... similar in scope to writing a survey paper on the area. (Neighbors, 1981)

Prieto-Diaz (1991) documents a method that he has used in several projects – the main inputs to the process are:

- technical literature
- existing implementations
- customer surveys
- expert advice
- current and future requirements

with outputs of:

- taxonomies
- standards such as interfaces
- functional models and generic architectures
- domain languages

A number of distinct roles are identified:

- domain expert   expert on area of application
- domain analyst   knows about domain analysis process/method
- domain engineer   implements domain models
- librarian   classifies assets, updates and distributes catalogue, tunes classification scheme
- asset manager   asset identification, acquisition, qualification, validation, and evolution
- reuse manager   reports on infrastructure performance

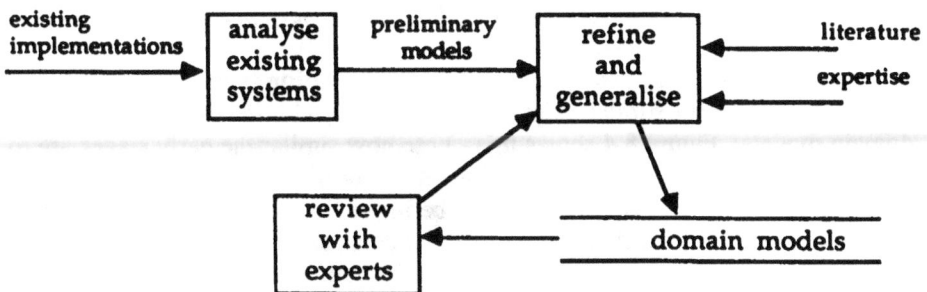

Figure 8.3 An example scenario for doing domain analysis

People playing these roles then 'do domain analysis', as illustrated in Figure 8.3. They:

- examine existing systems in the domain and other input materials
- develop generic architectures and domain models from systems descriptions

- identify difference between features, attributes, vocabularies, making trade-offs to create generic architecture (domain model)
- review generic architectures with domain experts which can then lead on to the identification of reusable assets and the deployment of these in future systems.

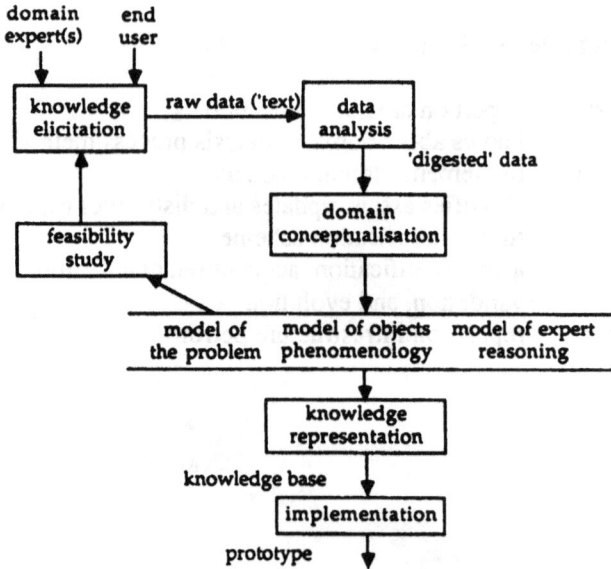

Figure 8.4 A knowledge acquisition method (Adapted from Motta et al., 1990)

If we look at methods used for knowledge elicitation, such as Motta et al. (1990), we find very similar stages in the process, with iterations to refine the domain models. Figure 8.4 shows this - note how similar the early stages are to the previous figure. Knowledge elicitation is not directly represented because our main source of knowledge is seen to be in previous systems - our 'raw text' which we analyse by hand. In Figure 8.4 the data analysis may use a number of tools, perhaps simple test analysis tools which count word frequencies, and find words which occur together ('collocation'). More sophisticated discourse analysis methods could also be used (Potter and Wetherall, 1986). The 'digested data' is then our preliminary model, with the generalisation and refinement being done here as domain conceptualisation. The feasibility study is our review by experts, while all the activities after the model has been established is component engineering. Not that the models are more explicit and apparently richer but, in domain analysis, any model or language necessarily comes with a set of rules of inference or reasoning

Arango has suggested that this approach is too optimistic, writing that domain analysis is:

... essentially difficult ... against the possibility of practical procedures. (Arango 1989)

He likens domain analysis to the creation of a scientific theory, and the pursuit of a domain analysis method as equivalent to the attempt to formalise scientific method, and hence doomed to failure. Nevertheless he recognises that there could be 'practical' procedures which could yield domain models adequate for a particular purpose.

This argument is persuasive, and we clearly should expect to depend on expertise and experience for domain analysis work, though this expertise could be focused and directed through a pragmatic method.

8.5 CONCLUSIONS AND RESEARCH DIRECTIONS

The software reuse position for domain analysis then appears to be:

Objective – produce a domain model which covers a number of related problems, but avoiding spreading the net too wide.

Representation – choose a representation scheme compatible with reusable assets to be deployed using the domain model. This is very likely going to be some model with an object-orientation flavour.

Method – we are going to have to rely on expertise, though we may use this within a framework or practical method. The methods of object-oriented analysis and knowledge elicitation provide a basis for detailed methods.

However we are left with a large number of open questions, such as:

- What defines the boundaries and scope of a domain?
- What could be a sufficient set of knowledge sources for domain analysis?
- How do we estimate the effort required to analyse a domain?
- How can we use reverse-engineering tools for domain analysis?
- Could we use natural language processing techniques?
- How far could we automate the classification and clustering of domain concepts?
- How could we handle viewpoints within domain analysis, and reconcile these into a single model?
- Would formal notations and methods help us?
- How does the domain representation chosen affect the ability to evolve the model?
- What skills and training or education is necessary for domain analysts?
- Are metrics applicable to domain models?

Domain analysis is clearly of much wider interest than software reuse. We see a common concern within the many threads in software development, with domain analysis emerging in:

- knowledge elicitation in artificial intelligence
- preservation of expertise, in knowledge engineering and conventional software engineering
- generalisation of requirements analysis
- description of an area of application in an IT strategy where a number of inter-related applications can be defined and evolved
- guide to decisions about the components to be placed in a reuse repository
- guide to the abstraction of descriptions in reverse engineering.

Many of these areas have their own special representations, methods, and tools, and we should now look a synthesis of these, and particular of knowledge engineering and software engineering. We need to exploit the power of knowledge based approaches in the service of more 'conventional' systems - and domain analysis gives us the opportunity to undertake that synthesis.

REFERENCES AND BIBLIOGRAPHY

Aitchison, J. and Gilchrist A. 1971, Thesaurus Construction. A practical manual, Aslib 1971.

Allen, J. 1987, *Natural Language Understanding*, Benjamin Cummings.

Arango, G. 1989, Domain Analysis – from art form to engineering discipline, Proceedings of 5th International Workshop on Software Specification and Design, pp. 152-159. Reprinted in Prieto-Diaz and Arango, 1991.

Atkinson, M., Bancilhon, F., DeWitt, D., Dittrich, K., Maier, D. and Zdonik, S. 1989, The Object Oriented Database System Manifesto, Rapport Technique Altair 30-89, 21 August.

Benyon, D.R. 1990, *Information and Data Modelling*. Blackwell Scientific, Oxford.

Bloomfield, B. 1986, Epistemology for Knowledge Engineers, CCAI 3, (4), pp. 305-320.

Boar, B. 1984, *Application Prototyping*, Addison-Wesley.

Booch, G. 1987, *Software Components with Ada. Structures, Tools and Subsystems*. Benjamin Cummings.

Brackman, R.J. and Schmolze, J.G. 1985, An overview of the KL-ONE Knowledge Representation System, *Cognitive Science*, **9** (2), pp. 171-216.

Breuker, J. and Wielinga, B. 1988, Models of Expertise in Knowledge Acquisition, Memorandum 103 VF-project Acquisition of Expertise.

Carasik, R.P., Johnson, S.M., Patterson, D.A. and Von Glahn, G.A. 1990, Towards a Domain Description Grammar: an application of linguistic semantics, ACM SIGSOFT Software Engineering Notes, **15** (5), October, pp. 28-43.

Clancey, W.J. 1983, The Epistemology of a Rule-Based Expert System, A Framework for Explanation, *Artificial Intelligence*, No 20, pp. 215-251.

Clancey, W.J. 1985, Heuristic Classification, *Artificial Intelligence*, No 27, pp. 289-350.

Cox, B. 1986, *Object Oriented Programming. An Evolutionary Approach*, Addison-Wesley.

Damiani and Bottarelli 1989, A Terminological Approach to Business Domain Modelling, Proceedings of IJCAI 1989. DARPA 1990, Proceedings of the Workshop on Domain Specific Software Architectures, July.

DARPA, 1988, Proceedings of the Case-based Reasoning Workshop, Darpa.

DARPA 1991, Proceeding of the Workshop on Informal Computing, July.

Devanbu, P., Brachman, R.J., Selfridge, P.G. and Ballard, B.W. 1991, Lassie: A Knowledge-Based Software Information System, CACM, **34** (5), May, pp. 34-49.

Dusink, L. and Hall, P.A.V. 1991, Software Reuse, Utrecht 1989, BCS workshop series, Springer Verlag.

Duverger, L., Van Damme, P. 1990, The Kake Project: A framework for the development of expert system projects based on modelling knowledge. KAKE internal report 1, BIKIT, Gent.

Floyd, C. 1987, Outline of a Paradigm Change in Software Engineering, Chapter 9 in *Computer and Democracy, A Scandinavian Challenge*, (eds.) Bjerknes Ehn and Kyng, Dower Publishing Company, England.

Forbus K.D. 1984, Qualitative Process Theory, Special Volume on Qualitative Reasoning about Physical Systems, Artificial Intelligence, **24** (103) pp. 85-169.

Freeman, P. (ed.) 1987, Software Reusability. IEEE Computer Society Press.

Gilchrist, A. 1971, The Thesaurus in Retrieval, Aslib, 1971.

Goguen, J.A. 1986, Reusing and Interconnecting Software Components, IEEE Computer **19** (2) February, pp. 16-28.

Grindley, K. 1975, *Systematics. A new approach to Systems Analysis*, McGraw Hill.

Hall, P.A.V. 1984, Relations, Logic, and Functional Programming, ACM SIG-MOD conference, Boston, USA, June.

Hall, P.A.V. 1988, Software components and reuse – getting more out of your code, in Information and Software Technology, Butterworths, Jan/Feb 1987. Reprinted in IEEE Tutorial, Software Reuse: Emerging Technology, (ed.) Tracz,W. IEEE Computer Society.

Hall, P.A.V. and Weedon R. 1991, Towards and Algebra of Objects, in preparation Hammersley and Atkinson, (1983). *Ethnography: Principles and Practice*, Routledge (originally Tavistock).

Hall, P.A.V. Boldyreff, C., Elzer, P., Keilmann, J., Olsen, L. and Witt, J. 1990, Practitioner: Pragmatic Support for the Reuse of Concepts in Existing Software, with ESPRIT, Week conference, Brussels, November 1990, Kluwer, Netherlands.

Harandi, M.T. and Ning, J.Q. 1990, Knowledge-Based Program Analysis. IEEE Software, January.

Harmon, P. and Sawyer, P. 1990, *Creating Expert Systems*, Wiley.

Hayward, S.A. Wielinga, B.J. and Breuker, J.A. 1987, Structured Analysis of Knowledge, *International Journal of Man-Machine Studies*, 26, pp. 487-498.

Horn W.(1991): The Challenge of Deep Models, Inference Structure and Abstract Tasks, Applied Artificial Intelligence, 5(1)47-56.

ISO 2788-1986; Documentation – Guidelines for establishment and development of monolingual thesauri.

ISO 5964-1985; Documentation – Guidelines for establishment and development of multilingual thesauri.

Jackson, M.A. 1983, *System Development*, Prentice Hall.

Keravnou, E.T. and Johnson, L. 1986, *Competent Expert Systems*, Kogan Page.

Keravnou, E.T. and Washbrook, J. 1989, What is a Deep Expert System? An analysis of the architectural requriements of second-generation expert systems. *Knowledge Engineering Review*, **4** (3) pp. 205-233.

Kim, W. and Lochovsky, F.H. (eds.) 1989, *Object-Orientated Concepts, Databases, and Applications*, ACM Press / Addison-Wesley.

Klein, H.K. and Hirscheim, R.A. 1987, A Comparative framework of data modelling paradigms and approaches, *Computer Journal*, **30**, 1, pp. 8-15.

Kolodner, J.L. (ed.) 1989, Proceedings of the Case-based Reasoning Workshop, Darpa.

Kolodner, J.K. 1991, Improving Human Decision Making through Case-Based Decision Aiding, *Artificial Intelligence*, **12**, (2), pp. 52-68.

Leith, P. 1986, Fundamental errors in legal logic programming, *Computer Journal* **29**, (6).

Leith, P. 1990, *Formalism in HIC Computer Science*, Ellis Horwood.

Maarek, Y.S. 1990, Indexing Software Components for Reuse by Using Natural-Language Documentation, position paper for 3rd annual workshop: Methods and Tools for Reuse, Syracuse, June.

Meyer, B. 1988, *Object-oriented Software Construction*, Prentice-Hall.

Morik, K. 1991, Underlying Assumptions of Knowledge Acquisition and Machine Learning, *Knowledge Acquisition* 3(2), pp. 137-156.

Motta, E. Rajan, T. and Eisenstadt, M. 1990, Knowledge acquisition as a process of model refinement, **Knowledge Acquisition, 2**, pp. 21-49.

Mumford, E. Land, F. and Hawgood, J. 1987, A participative approach to the design of Computer Systems, in Gallies, R. (ed.), *Information Analysis – Selected Readings*, Addison-Wesley.

Mumford, E. and Henshall, D. 1979, A participative approach to computer systems design, Associated Business Press.

Murray, F. and Woolgar, S. (eds.), in press, *Social Perspectives on Software*, MIT Press.

Neale, I.M. 1988, First generation expert systems: a review of knowledge acquisition methodologies, *Knowledge Engineering Review*, 3, (2), June, pp. 105-145.

Neighbors, J. 1984, The Draco Approach to Constructing Software from Reusable Components, in IEEE 84.

Nierstrasz, O. 1989, A Survey of Object-Oriented Concepts, Chapter 1 in Kim and Lochovsky.

Nijssen, G.M. (ed.) 1976, Modelling in Data Base Management Systems, North Holland.

Potter, J. and Wetherell, M. 1978, *Discourse and Social Psychology*, Sage.

Prieto-Díaz, R. 1990, Domain Analysis: An Introduction, Software Engineering Notes, 15, (2), April 1990.

Prieto-Díaz, R. and Arango, G. (eds.) 1991, Domain Analysis and Software System Modelling, IEEE.

Rau, L.F. 1987, Knowledge Organisation and Access in a Conceptual Information System, *Information and Management*, 23, (4), pp. 269-283.

Robinson, H.M. 1991, Reality and other practical problems of database engineering, Nizke Tatry, November 1991.

Rogers, Y. and Rutherford, A. 1991, (in press), Future directions for mental models research, in Rogers, Y., Rutherford, A., and Bibby, P., (eds.) *Models in the Mind: Theory, perspectives and application*, London, Academic Press.

Schreiber, G., Breuker, J., Bredeweg, B., Wielinga, B.J. 1988, Modelling in KBS development, in Boose, J.H. et al. (eds), Proceedings of the European Knowledge Acquisition Workshop (EKAW88), GMD-Studien Nr143, St Augustin, FRG.

Shaw, M.L. and Woodward, J.B. 1990, Modelling expert knowledge, *Knowledge Acquisition*, 2, 179-206.

Silverman, D. 1985, *Qualitative Methodology and Sociology*, Gower.

Simos, M.A. 1988, position paper for the Workshop on Tools and Environments for Reuse , Bass Harbor, Main, June.

Steels, L, 1990, Components of Expertise, Artificial Intelligence, **11** (2) pp. 28-49.

Stein, L.A., Lieberman, H. and Ungar, D. 1989, A Shared View of Sharing: The Treaty of Orlando, Chapter 3 in Kim and Lochovsky.

Suchman, L. 1987, *Plans and Situated Actions*, CUP.

Townley, H.M. and Gee, R.D. 1980, *Thesaurus Making. Grow your own wordstock*, Andre Deutsch.

Tracz, W. (ed.) 1988, Software Reuse: emerging technology, IEEE Computer Society Press.

Trost, H., Pfahringer, B. 1988, VIE-KL: An Experiment in Hybrid Knowledge Representation, Oesterreichisches Forschungsinstitut für Artificial Intelligence, Wien, TR-88-8.

Washbrook, J. and Keravnou, E.T. 1990, Making Deepness Explicit, *Artificial Intelligence in Medicine*, **2** (3), pp. 129-134.

Waters, R.C. 1988, Program Translation via Abstraction and Reimplementation, IEEE Transactions on Software Engineering, **14**, (8), August, pp. 1207-1228.

Woods, W.A. 1986, Important Issues in Knowledge Representation, Proceedings of the IEEE **74**, (109), pp. 1322-1334.

Woodward, B. 1990, Knowledge Acquisition at the front end: defining the Domain. Knowledge Acquisition, **2** (1) pp. 73-94.

Woolgar, S. and Lynch, M. (eds.) 1990, *Representation in Scientific Practice*, MIT Press.

Yourdon, E. 1989, Modern Structured Analysis, Prentice-Hall.

9 Application Templates: Reusable Design

Dr P. McParland
Institute of Software Engineering

For years developers have been reusing program code. They either extract sub-routines from a subroutine library or 'cut and paste' chunks of code from one application to another. However, the average developer still prefers to write his/her own program code. This is because:

- isolating reusable code in existing applications, or even in subroutine libraries, is difficult;
- developers write code to solve a particular problem. They do not concern themselves with the possibility of other developers wishing to reuse their code. Thus the work that a developer must undertake to make a piece of existing code reusable may make it simpler to develop the code from scratch.

Obviously, encouraging reusability at the code level presents major technical and cultural difficulties to an organisation.

Research over the last ten years has identified a number of possible techniques for encouraging reusability. Figure 9.1 (taken from Biggerstaff, 1984) describes the main approaches which are broken down into reuse of building blocks (such as subroutines or Objects in Object Orientation Technology) and the reuse of patterns or designs. Application templates fit into this latter category.

Features	Approaches to Reusability			
Component Reused	Building Blocks		[Patterns]	
Emphasis	Libraries	Composition	Language [Based] Generators	Application Generators
Typical Systems	NAG Subroutines	OO Pipes	VHLLs	4GLs CASE

Figure 9.1 A framework of reusability technologies

Application templates are generic specifications for specific types of application (e.g. financial, stock control and sales packages). They make use of the idea that many of the applications within such fields have a 60% or 70% overlap in data and functionality. The template provides the skeleton data models and processes for these packages and the developer tailors them for a specific application. However, application templates are more than just skeleton specifications. They consist of the analysis, the design and the low level code for a working application. That application is generic so it can form the foundation of a more detailed application.

For example, most stock control systems perform virtually the same generic tasks. They differ mainly in the type of stock they store and how the stock control system interfaces with other applications. An application template that defines a generic stock control system would be a working application. However, it would lack specific details on the nature of the items which the system stores (e.g. validation logic to check product numbers and specialised reports). Also, if the required stock control system is to interface with a bar code reader in a shop (say) then the developer may have to modify the template to incorporate such a relationship.

9.1 ADVANTAGES OF APPLICATION TEMPLATES

The most obvious benefit of using an application template to guide the development of an application is that developers' productivity should improve substantially. Instead of developing new specifications and code they are reusing an existing specification and its code. However, as well as increasing productivity, the use of application templates should also improve the quality of the developed application. Experts develop a template by extracting a generic application from many existing applications in the same field. They also test each template rigorously to validate it. Thus libraries of application templates may become invaluable to large organisations as a way of imposing quality and consistency on their applications development.

Currently, a company building an application (e.g. a stock control system), which they wish to use in different industries (e.g. a shop or a factory), provide such flexibility by producing highly parameterised code. Building in such flexibility at the code level is a major task. Using an application template, developers can have a different version of virtually the same application for each industry. The template will define the core facilities of the application. Developers can then specify each subsequent version of the application to suit the requirements of a specific client industry. Thus using application templates developers can create applications more in tune with the individual requirements of a client or with their own business.

The use of application templates also offers advantages over buying third party applications. The problem with using third party applications is that the organisation surrenders control of the application's future development to its supplier. However, an application, based on an application template, can evolve with an organisation since the organisation's developers can tailor the template's design to suit the changing needs of the business. Thus the use of an application template offers a more flexible solution to the provision of applications than is available from a third party package.

Therefore the use of an application template offers a combination of the benefits of buying a third party application with the flexibility and control associated with building your own application.

9.2 APPLICATION TEMPLATES AND CASE

To gain the most benefit from using application templates it is best to adopt them within the context of a CASE (Computer Aided Software Engineering) tool. A CASE tool supports a software development method that guides the developer through all stages of an application's development, from planning through to construction. If the CASE tool provides a code generator to automate the construction phase then the CASE tool can significantly increase productivity. Using a code generator also ensures that the CASE tool maintains a dynamic relationship between the application's code and its specification (see Figure 9.2). Developers make all updates to the application via the CASE tool's analysis and design tools and then re-generate the application's code using the code generator. Thus the code is always up to date with respect to its specification.

Many CASE vendors provide application templates for their CASE products. These templates take the form of pre-populated repositories containing all the analysis and design information necessary to allow a first-cut application to be generated. Developers can then use the CASE tool's analysis and design tools to amend the application template to suit their more specific needs. Such amendments can be made safely since the CASE tool ensures that the changes to the template are consistent and it generates the working application. Thus the developers would spend most of their time modifying the application's specification rather than writing program code.

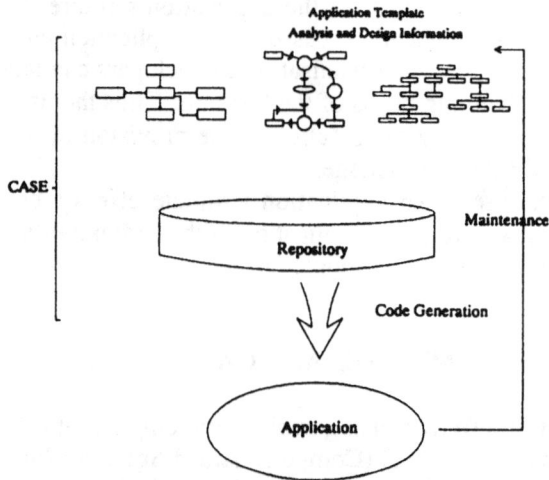

Figure 9.2 A typical CASE tool

By associating an application template with CASE tools, vendors are encouraging developers to view the application template as more than a paper-based specification. If the template is just a series of documents and program code then developers may ignore vital parts of the specification. Also, it is much more difficult to change the specification and then to replicate that change in the code if the development environment is not CASE based. For this reason application templates are often provided for a particular CASE tool.

9.3 WHO WOULD SUPPLY APPLICATION TEMPLATES?

A number of CASE vendors are marketing such templates currently (McParland, 1991). For example, Texas Instruments provide a General Ledger application template for IEF and Andersen Consulting provide templates (or Designware products) for life insurance systems, customer information systems and others.

9.4 AN EXAMPLE APPLICATION TEMPLATE FROM ANDERSEN CONSULTING

As mentioned above, Andersen Consulting provide several application templates (or Designware products) for users of their CASE tool FOUNDATION. One of their application templates is called CUSTOMER/1. CUSTOMER/1 is a blueprint for developing a complete customer in formation system. It handles the entire customer contact life cycle and provides tracking of relationships between customers, accounts, products and services.

Andersen Consulting developed the original version of CUSTOMER/1 for the Baltimore Gas and Electric Company in 1985 and it has since been generalised for use in many companies. Thus CUSTOMER/1 is not an untried design available to FOUNDATION users but it is the design and code of a comprehensive product which is used in the utilities industry throughout the world.

CUSTOMER/1 provides developers with a complete modifiable customer information system which includes 350 screens, 200 reports and over 2,000 data elements. All of this design information is available to the developer via the FOUNDATION CASE tool. Developers can customise this design and generate a working application suited to their company's individual requirements.

CUSTOMER/1 consists of a wide range of subsystems, or high level functions. These are:

- the Service Order subsystem which processes orders to initiate, change or terminate metered services provided to customers;
- the Meter Reading subsystem;
- the Billing subsystem;
- the General Data Maintenance subsystem to update data associated with customers and their accounts (e.g. changes of name, etc.);
- the Accounts Data Maintenance subsystem which processes all financial transactions associated with the customer;
- the Credit and Collection Subsystem which assists with credit controls;
- the Gross Functional subsystem which is a collection of functions to define general purpose processes, procedures and rules.

Space constraints preclude the description of each of these subsystems but the Service Order subsystem is presented in more detail.

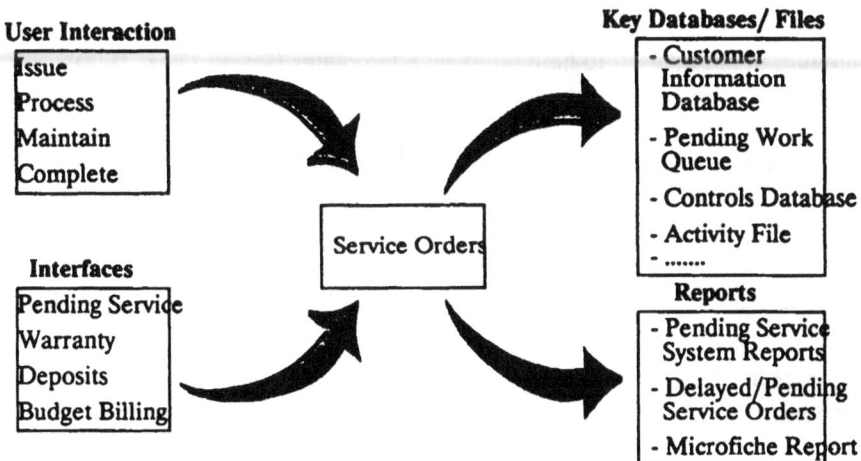

User Interaction

Issue
Process
Maintain
Complete

Interfaces

Pending Service
Warranty
Deposits
Budget Billing

Service Orders

Key Databases/ Files

- Customer Information Database
- Pending Work Queue
- Controls Database
- Activity File
- ....

Reports

- Pending Service System Reports
- Delayed/Pending Service Orders
- Microfiche Report

Figure 9.3 The Service Order subsystem

The Service Order subsystem consists of a single function which processes orders to initiate, change or terminate metered service provided to customers. The function is broken into three steps:

1. A customer requests the order.
2. A work order is generated for field completion.
3. The order is completed and the Customer Information database is updated.

The Service Order design includes order screens for several types of gas and electric services:

- Turn On Service
- Turn Off Service
- Turn On/Off Incontinuation
- Change Meter
- Remove Meter
- Set Meter
- Install Private Area Lighting.

Andersen Consulting can provide further information on the details of CUSTOMER/1 and their other Designware products.

9.5 SUMMARY

The above example shows how CASE vendors are providing templates for their customers to use as blueprints for some of their applications. However, it need not be only CASE vendors who market such templates. New companies could emerge which would specialise in selling libraries of templates. Domain analysis techniques could be used to help build such templates (Prieto-Diaz, 1990).

Many DP departments may already have all the templates they need hidden in their existing applications. In many business applications the data is relatively stable so there may be valuable assets among those old COBOL listings! However, extracting the generic templates from existing applications (even with reverse engineering tools or Domain Analysis) could be a daunting task.

REFERENCES

Biggerstaff, T.J. 1984, Foreword to IEEE Transcription on Software Engineering, SE-10, No.5, September.

McParland, P. 1991, Application Generation: Automated Software Development Using 4GEs and CASE, Institute of Software Engineering Report (Tel. +44 232 738507).

Prieto-Díaz, R. 1990, Domain Analysis: An Introduction, ACM SIGSOFT Software Engineering Notes **15**, (2).

10 Human Issues in Software Reuse

Neil Maiden
City University

10.1 INTRODUCTION

Software reuse has been proposed to improve the quality and productivity of software development for over 30 years, however the widespread uptake of software reuse has failed to materialise, for two possible reasons. First, software managers failed to invest in software reuse since its benefits were either intangible or unquantified. Second, technical solutions for the retrieval of reusable software were seen to be inadequate for widespread reuse. One potential reason for this is that most technical solutions fail to recognise the need and potential of human involvement in reuse. Indeed, the human role has received little attention in the literature, so this paper investigates specific problems which arise from excluding software developers in the reuse process, and proposes potential solutions based on human involvement for facilitating and promoting widespread reuse.

Early approaches to software reuse exploited low-level code modules during system construction. Modules were treated as black boxes from which systems were composed using techniques including parametrisation and modular interconnection languages, and the programmer was not required to inspect or modify code within these modules. Indeed, module adaptation was deemed to lessen the inbuilt quality of the reusable product. However, CASE (Computer-Aided Software Engineering) technology supporting the more critical analysis and design phases of software development necessitates reuse of larger, application-specific components which require good understanding for effective customisation. Adaptation of reusable software, designs and specifications is a complex, knowledge-intensive task requiring extensive involvement of the software developer. However little is known of the problems involved in this task, so the processes of software comprehension and customisation were investigated.

Empirical studies of software reuse, system maintenance and program debugging suggest that effectively understanding unfamiliar software is both difficult and time-consuming, even for expert software engineers. Expert programmers require considerable time and mental effort to successfully debug unfamiliar programs while novice programmers fail to understand either the program function or structure (Holt et al., 1987). Indeed novices tended to adopt strategies which hindered understanding (Nanja and Cook, 1987). Understanding during software reuse has also proved problematic. Langer and Moher (1989) reported studies of object-oriented programming which suggested that mental laziness was manifest, as copying during reuse was employed as one mechanism to compensate for an incomplete mental model rather than as a shortcut method when the mental model was mature. Investigations of problem-solving in other domains have yielded similar results; for instance Chi et al. (1989) reported that students incorrectly solved physics problems by copying example solutions with shared salient properties, while experts correctly categorised and solved problems from the underlying physics law. In light of these findings, software reuse researchers are now becoming aware of the human-oriented problems which inhibit software understanding and adaptation (Thompson and Huff, 1991). This paper examines the nature of these problems in terms of some typical reuse examples then proposes several solutions intended to facilitate effective reuse.

10.2 EXISTING PROBLEMS WITH SOFTWARE REUSE

Human involvement is needed in the following software reuse tasks:

i) describing new problems in terms which permit the retrieval of reusable components;
ii) determining the appropriateness of a retrieved reusable component for solving a new problem;
iii) selecting the best component from several candidate reusable components;
iv) adapting a selected reusable component to fit the new domain;
v) adding extra functionality to the component while ensuring the validity of the existing component features.

In short, human-related problems during reuse can be broken down into three basic types. First, software engineers must define new problems using constrained terminology. Second, retrieved reusable components must be effectively understood, otherwise they may be reused inappropriately or incorrectly, and third, understood components must be effectively transferred and adapted to fit the new domain and maximise the potential reuse. To demonstrate these problems, each is examined in terms of one or several reuse examples which highlight the potential pitfalls for software reusers.

10.2.1 Defining new problems to be solved by reuse

Most software reuse paradigms require new system needs defined in terms of reusable components. They assume that software engineers can define complete and correct system needs using a constrained set of terms, so little tool-based support for this task has been provided. However, Furnas et al.'s (1987) study of vocabulary-driven interaction during design found developers' word choices for objects to be surprisingly large, suggesting inherent problems with lexically-based, keyword retrieval mechanisms. In short, software developers may be unable to select or describe new problems in terms of existing schemes. Consider the following example. Prieto-Diaz and Freeman's (1987) well-known software classification scheme requires software engineers to describe new problems in terms of six facets describing the Function, Objects, Medium, System Type, Functional Area and Setting. These facets represent program modules rather than domain-specific specifications or designs, so the current scheme is not complicated by application-specific terms or structures. However, for each of these facets there can be many instances, especially for Functional Area and Setting (see Table 10.1). Indeed, later work by Prieto-Diaz (1991) suggests that these classification schemes have been extended beyond the scheme shown in this chapter.

Function	Objects	Medium	System type	Functional Area	Setting
add	arguments	array	assembler	accounts payable	advertising
append	arrays	buffer	code generation	accounts receivable	appliance repair
close	backspaces	cards	code optimisation	analysis structural	appliance store
compare	blanks	disk	compiler	auditing	association
complement	buffers	file	DB management	batch job control	auto repair
compress	characters	keyboard	expression evaluator	billing	barbershop
create	descriptors	line	file handler	bookkeeping	broadcast station
decode	digits	list	hierarchical DB	budgeting	cable station
delete	directories	mouse	hybrid DB	capacity planning	car dealer
divide	expressions	printer	interpreter	CAD	catalog sales
evaluate	files	screen	lexical analyser	cost accounting	cemetery
exchange	functions	sensor	line editor	cost control	circulation
expand	instructions	stack	network DB	customer information	classified ads
format	integers	table	pattern matcher	DB analysis	cleaning
input	lines	tape	predictive parsing	DB design	clothing store
insert	lists	tree	relation DB	DB management	composition
join	macros	.	retriever	.	computer store
measure	pages	.	scheduler	.	.
modify	.	.	.	.	.
move	.	.	.	.	.
.	.	.	.	.	.

*Table 10.1 Partial listing of faceted classification scheme
(From Prieto-Diaz and Freeman 1987, p.10)*

Descriptor browsing is rendered complex by the number of facets, many of which may be candidate descriptors for the new problem. Understanding the precise meaning of each descriptor is also difficult, since simple facets can be understood by different people in different ways (Furnas et al., 1987).

Furthermore, many terms are semantically close and require differentiation to be effectively used (e.g. append, add and insert for Function), while use of the same terms for different facets is a further complication. Providing more detailed definitions of descriptors can help to overcome this comprehension problem, although it noticeably increases the amount of data to be assimilated by the software engineer during problem definition. Intelligent tool-based support which attempts to infer software engineers' intentions from their use of constrained terminology may partially alleviate this problem, however such lexicons or thesauri are difficult to derive, knowledge-intensive, and may exacerbate the already-difficult task of maintaining such schemes (Prieto-Diaz, 1991).

In short, description of new problems using existing terms and schemes may be more problematic than assumed in the software reuse literature. Potential solutions in the shape of powerful, generic and easily-understood problem specification languages are not readily available, so the problems encountered by humans when defining problems to be solved by reuse warrant further research. However, human problems also arise during comprehension and customisation of reusable components. A recent IBM study on software maintenance suggested that 50% of all maintenance time was spent understanding code modules to be changed, suggesting potential problems during code understanding for both maintenance and reuse. The remaining examples demonstrate these problems in terms of understanding and adapting retrieved reusable components.

10.2.2 Problems of component understanding

Problems inherent in component understanding are demonstrated by two simple reuse examples of small- and large-scale component understanding.

Example of small component reuse

The first example demonstrates difficulties encountered when determining the functionality and nature of unfamiliar but small code components. Consider the code components shown in Figures 10.1 and 10.2, and try to determine their purpose, structure and conditions for reuse. In these instances the component structure and variable-naming in both components can make this task difficult and time-consuming without prior knowledge of the components and the problem which they were originally intended to solve. Difficulties are exacerbated by the need to browse and assimilate many similar modules, which may be difficult even for expert software engineers. Furthermore, reuse may be even more difficult for inexperienced software engineers who potentially have most to gain from exploiting such expertise. Unfortunately successful reuse requires effective understanding, but understanding requires prior knowledge of the component, so lessening the payoff from reuse to individual software engineers.

```
imp Bseq ($$u)
rep {| struct $BSEQ {
        $*_seq[$$u];
        int _hd _tl;
}; |}
repty (seq ($*), Bseq) = {| struct $BSEQ |}
hd ($s) = {|
        $s._seq [$s._hd]
|}
isnil ($s) = {|
        ($s._hd == $s._tl)
|}
assign (Ss, nil()) = {|
        $s._hd = $s._tl = 0;
|}
assign ($s, tl (var $s)) = {|
        $s._hd++;
|}
assign ($s, apndr (var $s, $x)) = {|
        if ($s._tl == $$u)
        fputs ("bound $$s exceeded", stderr);
        else {
           assign ($s._seq[$s._tl], $x)
           $s._tl++;
        }
|}
assign ($ts, $s) = {|
        $ts._hd = $ts._tl = $s._hd;
        for ($ts._tl != $s.tl; $ts._tl++) {
           assign {$ts._seq {$ts._tl], $s._seq {$
        }
|}
```

Figure 10.1 'Bounded 'C Implementation of seq (*)', demonstrating the complexity involved in understanding unfamiliar software components

Example of large component reuse

The previous example demonstrated difficulties associated with understanding simple components. However, industrial scale reuse is likely to be achieved from much larger reusable components, for instance Kruzela (this volume) identified that successful reuse within the Japanese telecommunications company NTT was achieved from components with upward of 600 lines of code. Reuse of these larger components is likely to magnify problems identified in the previous example as well as introduce new problems associated with large-scale components:

- limitations of current program-editing technology do not encourage effective understanding of large software components. Many computer screens only show 20 lines of code at any time, making module browsing difficult. Older software development environments do not support multi-windowing facilities to assist this browsing, so obtaining a comprehensive overview of the code's functionality and structure from the program itself is difficult;
- most documentation associated with existing software (especially systems in commercial environments) is ambiguous and incomplete, so it cannot be relied upon to obtain an overview of large component functionality. Structured programming methodologies and modular program develop-

```
                          Page 1
with Text_IO;  use Text_IO;
procedure Sort_Names is
  subtype Names is String (1 .. 20);
  Current_Name : Names;

package Sort_Service is
  procedure Make_List_Empty;
  procedure Insert (New_Name : in Names);
  procedure Prepare_Extraction;
  procedure Extract (Old_Name : out Names);
  function All_Extracted return Boolean;
end Sort_Service;

package Name_IO is
  procedure Get_Name (Name : out Names);
  procedure Put_Name (Name : in Names);
end Name_IO;

use Sort_Service, Name_IO;

package body Sort_Service is
  Max_Size  : constant Positive := 100;
  List    : array (Positive range 1 .. Max_Size) of Names
  Size, Extracted: Natural range 0 .. Max Size;

  procedure Make_List_Empty is
  begin
   Size := 0;
  end Make_List_Empty;

  procedure Insert (New_Name : in Names) is
  begin
   insert New_Name in its correct position in the list
  end Insert;

  procedure Prepare_Extraction is
  begin
   Extracted := 0;
  end Prepare_Extraction;

  procedure Extract (Old_Name : out Names) is
  begin
   Extracted := Extracted + 1;
   Old_Name := List (Extracted);
  end Extract;
```

```
                          Page 2
  function All_Extracted return Boolean is
  begin
   return (Extracted = Size);
  end All_Extracted;
end Sort_Service;

package body Name_IO is
  procedure Get_Name (Name : out Name
  begin
   read Name from a single line of input;
  end Get_Name;

  procedure Put_Name (Name : in Name
  begin
   Put (Name);
  end Put_Name;
end Name_IO;

begin
  Make_List_Empty;
  while not End_of_File loop
   Get_Name (Current_Name); Skip.
   Insert (Current_Name);
  end loop;
  Prepare_Extraction;
  while not All_Extracted loop
   Extract (Current_Name);
   Put_Name (Current_Name); Nev
  end loop;
end Sort_Names;
```

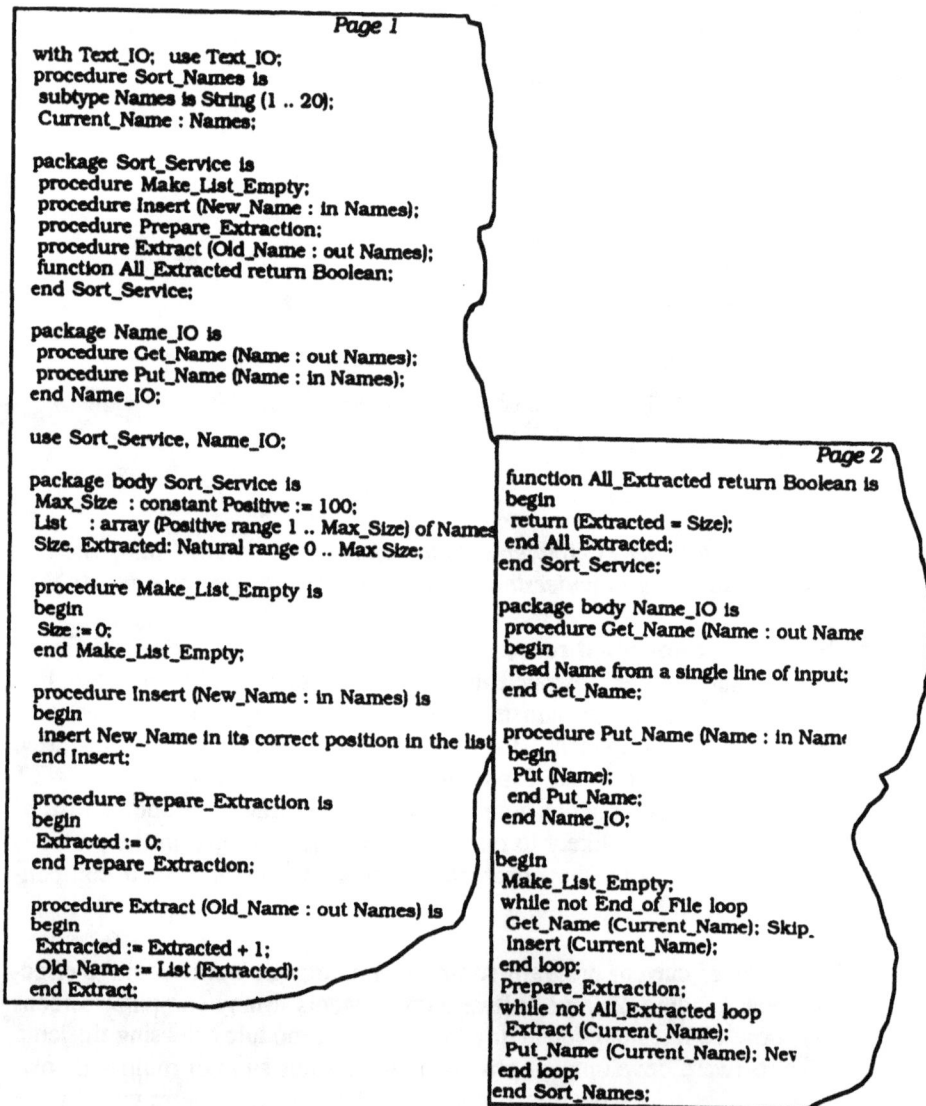

Figure 10.2 Ada module sorting a list of all names
(From Watt, Wichmann and Findlay, 1987)

ment can lead to the development of better documentation, but their uptake has been far from widespread;

- understanding large code modules is a cognitively-complex task, even if effective tool support and documentation are provided. Reuse of large-scale components also emphasises the importance of information assimilation and structuring. At the very least, understanding is likely to need electronic notepad facilities to record reasoning and assist module structuring (Haddley and Sommerville, 1990).

To summarise, understanding large and unfamiliar code components appears to be very difficult without tool-based support and guidance. Component understanding is necessary for effective component selection, however this selection process introduces additional factors which can further inhibit successful reuse.

Example of software component selection
The fourth example demonstrates the difficulties inherent in selecting the most appropriate code component from several, similar candidate component. Problems of comprehension are compounded by the need to identify and understand small but critical differences between each component. Identification of these critical differences may be difficult since they were only represented by several lines of code. In addition, detailed reasoning about critical differences by the software reuser is needed to determine the best fit with detailed requirements for the new problem. These additional selection difficulties are demonstrated by the following, simple example. A programmer is writing a system for staff payments using COBOL: the required program section should determine and make appropriate income tax deductions for each employee. Two candidate reusable modules have been retrieved for solving this problem under the classification *income tax calculations* (see Figure 10.3).

The critical difference between the two modules is the inclusion of a check for each employee's tax code to determine whether tax should be paid, however this condition is poorly-defined within the first component, so making correct component selection difficult. In addition, browsing a space of similar components is both difficult and cognitively demanding, for instance the programmer is required to identify and recall fundamental differences between components. Furthermore, determining components' critical features may be difficult, for example, syntactic similarities between reusable components can interfere with the understanding and differentiation processes. Although simple, these problems are typical of those encountered during component selection, especially in the reuse of larger, real-world software components.

```
Begin-File.

    Start Employee Key is ".
    Read  Employee Next At End Go To End-File.
    Move Salary to Calc-Salary.
    Compute (NI=Calc-Salary*0.08).

Check-Bracket.

    If Tax = 'H' Go To Other-Bracket.
    Compute (Net=Calc-Salary*0.33).
    Compute (Taxable=Calc-Salary*0.25.
    Move Net To Month-Net.
    Go To End-Bracket.

Other-Bracket.

    Compute (Net=Calc-Salary*0.48).
    Compute (Taxable=Calc-Salary*0.41).
    Move Net To Month-Net.

End-Bracket.

    Move NI to Employee-NI.
    Rewrite Employee From Employee-Record.

Next-Read.

    Read Employee Next At End Go To End-File.

End-File.

    Stop Run.
```

Module A

```
Begin-File.

    Start Employee Key is ".
    Read  Employee Next At End Go To End-File.
    Move Salary to Calc-Salary.
    Compute (NI=Calc-Salary*0.08).
    If Code='NT' Go To End-Bracket

Check-Bracket.

    If Tax = 'H' Go To Other-Bracket.
    Compute (Net=Calc-Salary*0.33).
    Compute (Taxable=Calc-Salary*0.25.
    Move Net To Month-Net.
    Go To End-Bracket.

Other-Bracket.

    Compute (Net=Calc-Salary*0.48).
    Compute (Taxable=Calc-Salary*0.41).
    Move Net To Month-Net.

End-Bracket.

    Move NI to Employee-NI.
    Rewrite Employee From Employee-Record.

Next-Read.

    Read Employee Next At End Go To End-File.

End-File.

    Stop Run.
```

Module B

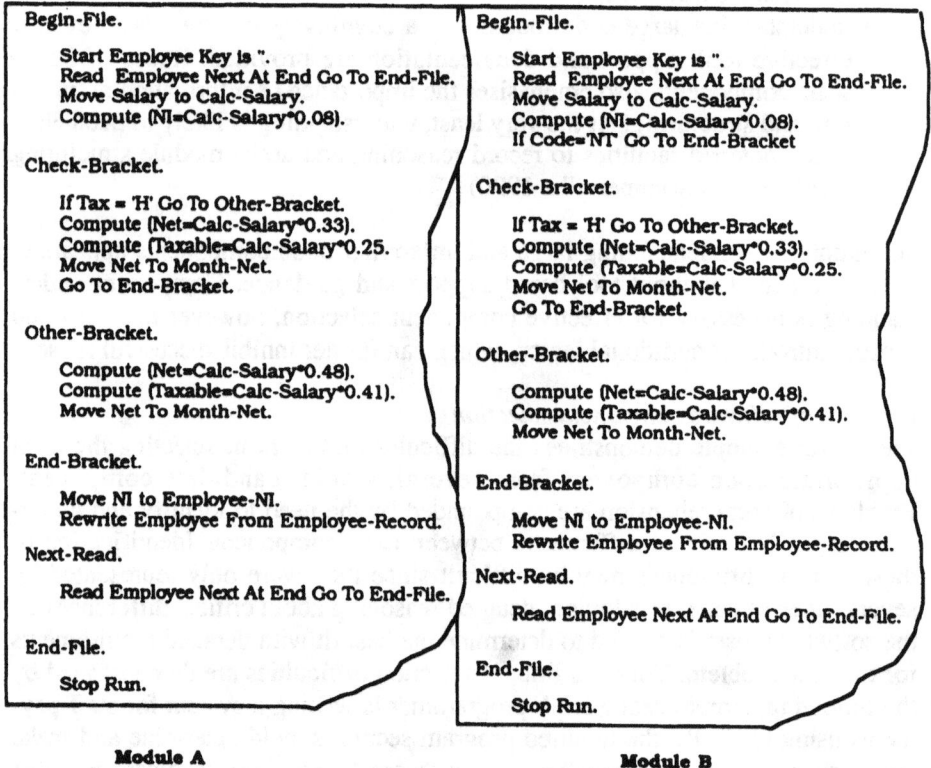

Figure 10.3 Two reusable COBOL modules representing two functions for calculating employee's income tax deductions

10.2.3 Problems of component adaptation

Adaptation and understanding of reusable components was examined in terms of two further examples, supported by empirical studies of reuse in terms of these examples.

Example of adaptation in software design

The first example is based on studies of software reusers' adaptation and comprehension strategies adopted during reuse of design-level components. It differs from previous examples since its main focus is on component customisation rather than understanding, and on reuse of system designs rather than lower-level code modules, although results can be applied to all levels of reuse. It is based on an empirical study reported more fully in Sutcliffe and Maiden (1990).

Thirty inexperienced software developers (MSc students in Business Systems Analysis) were asked to develop a JSD (Jackson, 1983) process structure diagram for a scheduling function in a video hiring company, in which video tapes were allocated to hotels if they met specified requirements. Two reusable designs were developed. A production planning system allocated manufacturing machines to production jobs while a generic scheduling function allocated resources to tasks which had to be fulfilled. The main concept with both the concrete and generic designs was the functional requirement to allocate a resource within certain constraints, as shown in Figure 10.4. The reusable designs as well as the reused solution to the video hiring problem are shown in Figures 10.5, 10.6 and 10.7.

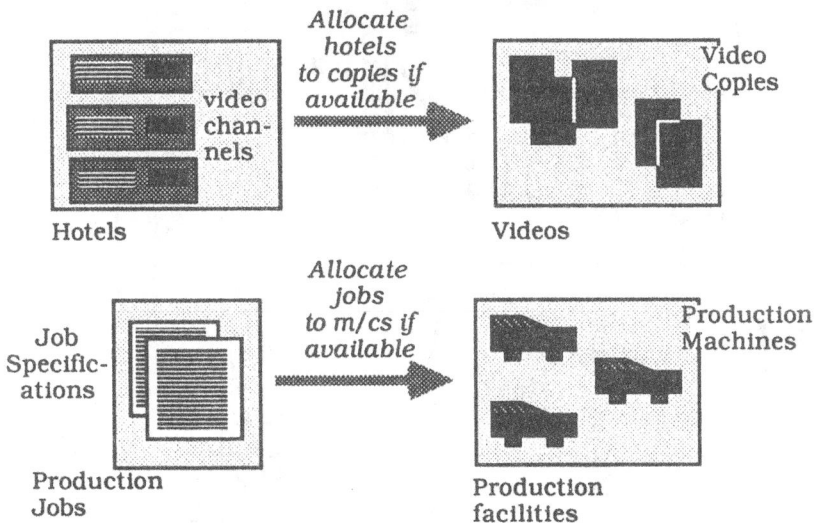

Figure 10.4 Domain models representing similarity supporting reuse in example of adaption in software design

The software engineers were divided into three groups to examine the effectiveness of reusing concrete and generic designs. Ten software engineers designed the video hiring function without any help, ten software engineers reused the concrete design and ten software engineers reused the generic design. Subsequently all design solutions were scored for completeness and validity by two expert software engineers with experience and knowledge of both applications. Results revealed that reuse significantly improved design completeness but not validity. Reusers were able to recognise and exploit the reusable designs despite evidence of poor understanding of the reusable designs. Mental laziness manifest as design copying in both reuse groups occurred, for instance nine software engineers correctly reused the JSD backtracking concept, although only three retrospectively admitted to understand its meaning. Furthermore, errors in the designs of 11/13 successful reusers were consistent with poor comprehension

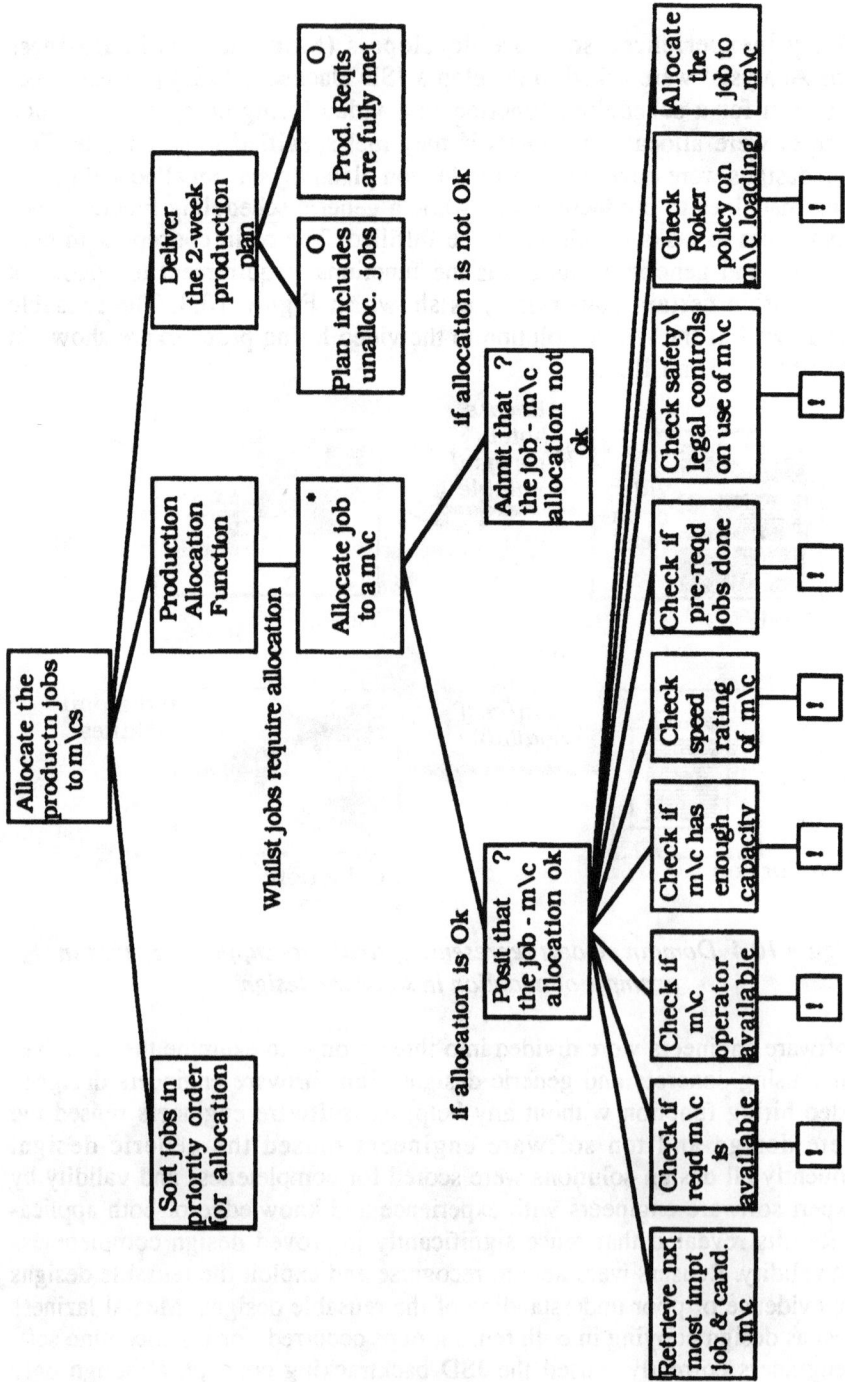

Figure 10.5 PSD for the overall production planning job-Mc allocation function

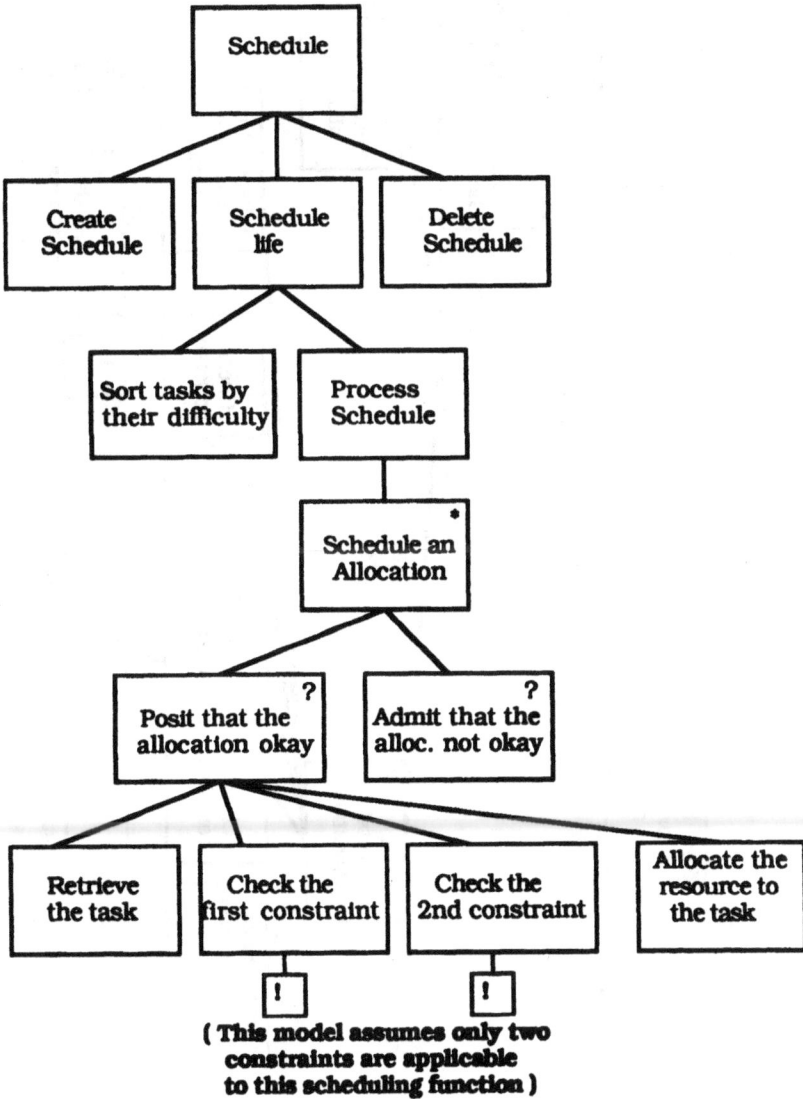

Figure 10.6 Template process structure diagrams for a scheduling function

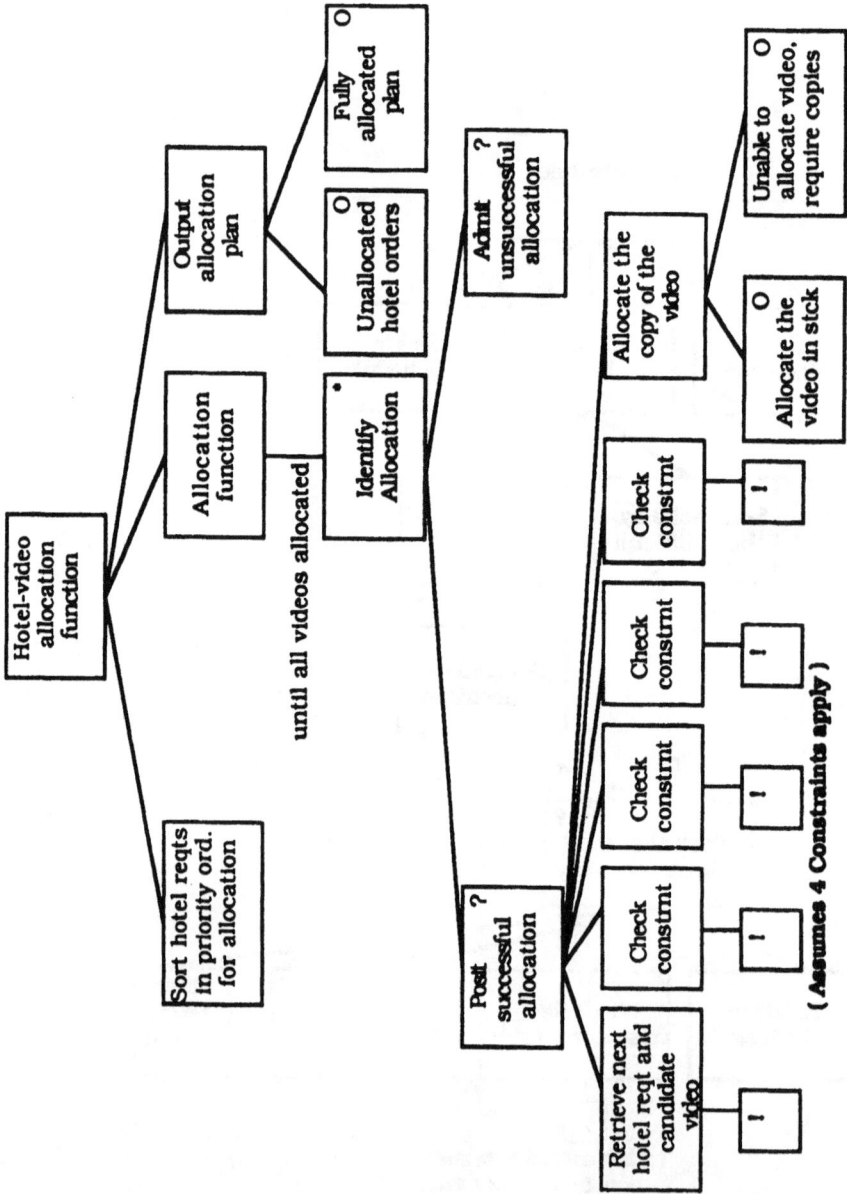

Figure 10.7 PSD for the VI allocation function

of the reusable designs. Software engineers exploited salient similarities between components in the target and reusable domains as a basis for reuse; for instance one software engineer incorrectly identified a similarity between production machine and video machines in hotels despite the invalidity of reusing the production machine to model the hotel video machines.

To conclude, this study revealed evidence of mental laziness and copying during software reuse, suggesting that it provided a shortcut to software engineers with only partial understanding of the components being reused, a result also found by Lange and Moher (1989) during program reuse. Transfer and customisation problems appear to arise then with poor component understanding, so the final example looked at these problems in greater detail.

Example of comprehension and adaptation during specification reuse

The final example also investigated problems of component comprehension and customisation. As with the previous example, findings are based on empirical studies, this time examining specification reuse during requirements analysis, although findings are also applicable to reuse during design and coding. Two empirical studies investigated analogical specification reuse, the first looking at reuse by five inexperienced software engineers and the second looking at reuse by ten expert software engineers, as described more fully in Maiden and Sutcliffe (in press, in preparation). In both studies software engineers were required to specify an air traffic control system by reusing an analogical flexible manufacturing system specification, see Figures 10.8 and 10.9. Reuse by justified by the shared concepts of object moving in space, risking collisions, following flight plans and being controlled by external agents. Protocol analysis (i.e. speaking out loud while thinking) was used to elicit software engineers' reasoning behaviour while video cameras captured other activities. All software engineers completed specifications which were scored for completeness and validity as in the previous study.

Results from the first study indicated that inexperienced software engineers copied the reusable specification. Ineffective reuse strategies (e.g. *partial reuse mixed with design of new components*) were identified and a set of malrules representing incorrect reuse were developed. Software engineers only understood objects or concepts which were syntactically similar and prominent in the reusable specification, at the expense of other reusable concepts. On the other hand, expert software engineers in the second study avoided copying and exhibited effective strategies which maximised transfer and adaptation of the specification and exploited all reusable components. However, understanding both the reusable specification and the analogy with it was difficult despite the best efforts of the experts, and like the novices, they only effectively understood analogical concepts which were syntactically similar and prominent in the reusable specification.

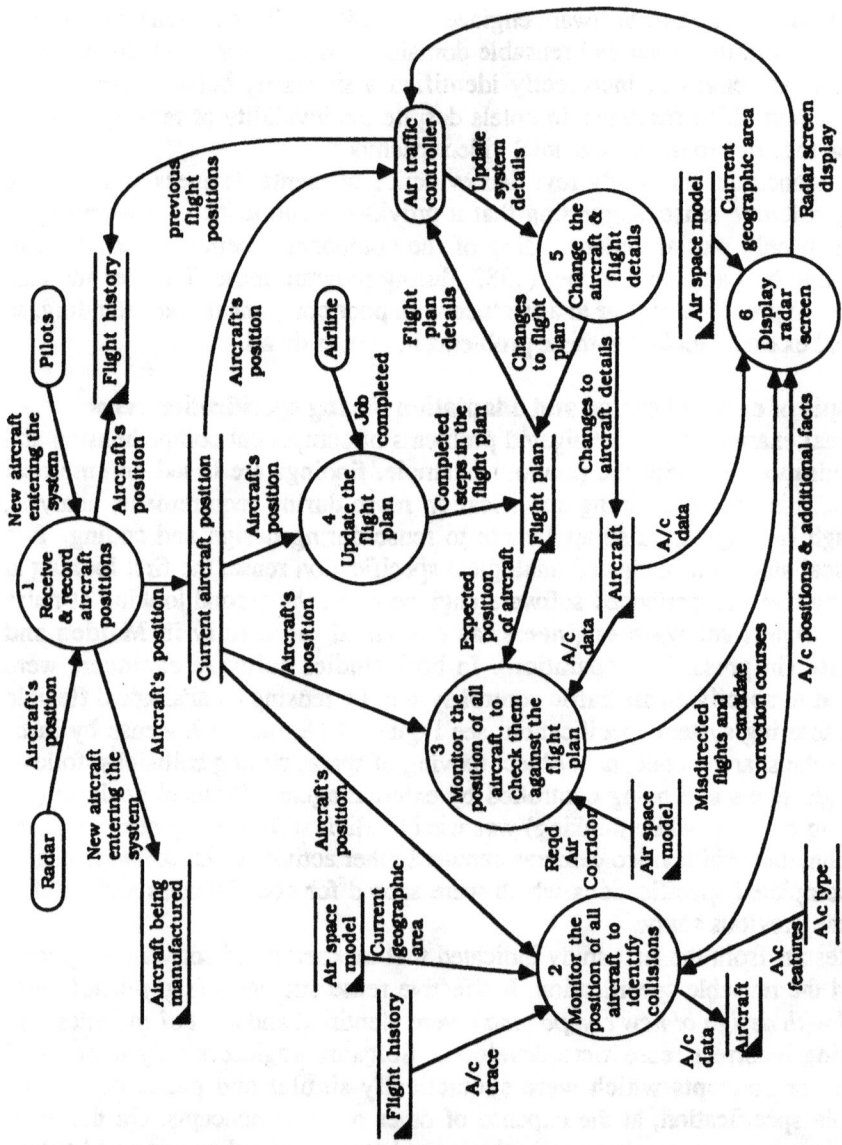

Figure 10.8 Level-0 DFD for Sunderland Air Traffic Control system

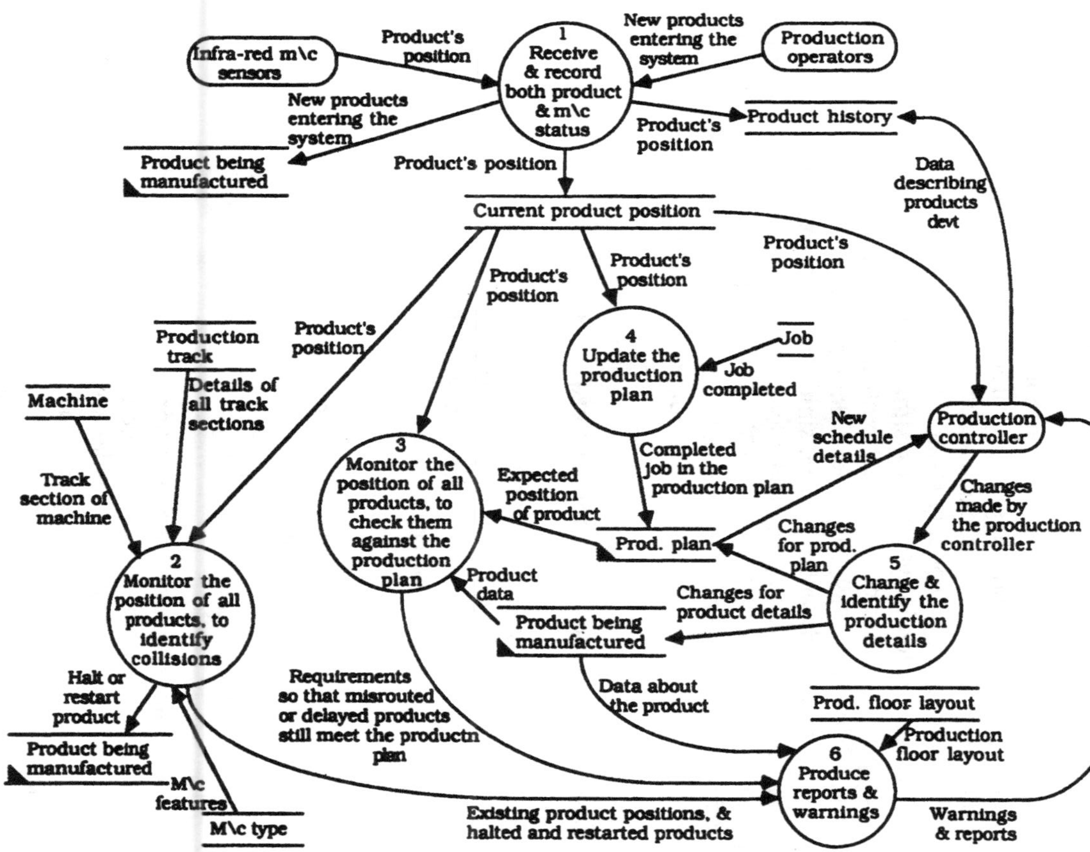

Figure 10.9 Level-0 DFD for Brockville Flexible Manufacuring system

These two studies revealed that, even with moderate-sized components represented using notations developed for ease of comprehension, understanding was difficult if software engineers were unfamiliar with the concepts or domains being reused. The emphasis on analogy in this final example demonstrates an additional complexity for software comprehension during reuse which does not arise during design, debugging or maintenance tasks. Effective reuse requires the software engineer to reason not only about the reusable component, but also the new problem and potentially-complex similarities and differences with it. Previous empirical studies of analogical problem solving (Gick and Holyoak, 1983; Gick, 1989; Ross, 1989; Novick and Holyoak, 1991) suggest that analogical recognition is difficult without salient similarities, schema induction is difficult without explicit prompting from additional sources, and analogical transfer does not automatically follow from effective analogical comprehension. Findings from the three studies reported in this paper concur with many of these conclusions, suggesting a greater need to focus on similarity-based reasoning during reuse.

10.2.4 Reuse problems arising from these examples

The six examples presented in this section can be interpreted either as a pessimistic conclusion for software reuse or as a challenge for the next generation of tools supporting industrial-scale reuse. They identify a number of critical, human-based issues worthy of consideration when developing software reuse environments:

- *the vocabulary problem when defining requirements for reuse*: the software engineer must be able to define system needs using terminology which the reuse retrieval mechanism can interpret correctly, i.e. the reuse system model must match the software engineer's situation model of the new problem;
- *component understanding*: identification of critical component functionality, structure and boundaries from complex, often undocumented, code components is a difficult and complex task;
- *data assimilation:* assimilating and understanding many large-scale code components which provide the greatest benefits from reuse is also problematic;
- *component selection*: identification of small but critical differences between retrieved components during selection of the best fitting module is necessary;
- *similarity-based reasoning*: the need to understand the reusable component and the similarities and differences between it and the new problem, akin to analogical problem solving is needed, especially during reuse of larger components;
- *the mental free-lunch*: the likelihood of mental laziness manifest as copy-

ing during customisation of poorly-understood components has been identified as a potential problem for software reuse throughout the software development life cycle.

Some of these problems have already been identified in the software reuse literature, others have often been ignored or down-played. The remainder of this paper examines tool-based support needed to overcome these problems in terms of knowledge requirements and tool functionality.

10.3 TOOL SUPPORT FOR EFFECTIVE SOFTWARE REUSE

Tool support is examined in terms of techniques to assist component comprehension and adaptation, and the knowledge which is available to assist these processes.

10.3.1 Knowledge to assist software reuse

Six knowledge sources can aid module understanding: original code documentation, reverse engineered code descriptions, design decisions or rationale underlying reusable components, explanations derived from criteria for module selection, comprehensive domain analysis, and cognitive modelling of software reusers. Each is examined in turn.

Original code documentation
Original code documentation can assist comprehension of code components. However, relying on original documentation may be problematic: software engineers are notoriously bad documenters, so code documentation may be incorrect, incomplete, ambiguous or even non-existent in many cases, especially if the system is old. Reliance on code documentation is only suggested if it can assuredly be said to be complete, accurate and up-to-date. Effective software reuse will need to rely on alternative knowledge sources in many instances.

Reverse engineering code descriptions
Current reverse engineering techniques have been proposed to extract specifications from code. One so far largely overlooked role for these reverse-engineered specifications is to assist understanding of reusable components. Even simple, syntactic-based reverse engineering techniques may prove successful in deriving major software functions and structure. Several reverse engineering approaches have recently been put forward in the literature, however, ironically their success is dependent upon extensive human involvement in the reverse engineering process. A recent study of reverse engineering (Byrne, 1991) revealed important issues about recovering design information, such as separating design details from implementation details, dealing with incomplete or erroneous information,

traceability of information between implementation and recovered design, and re-engineering. Further similar studies appear necessary to determine what knowledge may be effectively retrieved. However, even simple reverse-engineered software descriptions may assist reusable component comprehension, and the role of these descriptions may be promoted by reports of several success stories in the comprehension of reusable software.

Capture design decisions underlying design/code components

Design decisions and rationale underlying the development of reusable components can provide justification for their structure and implementation. Unfortunately, as with code documentation, rationale underlying many existing components will not be available, but capturing design decisions underlying new components can provide a basis for their explanation during comprehension. Initial research examined automatic replay of design rationale in well-understood domains (Mostow, 1989). However, several techniques currently exist for capturing such decisions; for instance Ibis (McCall, 1992) uses question/answer techniques to obtain important decisions behind reusable components, while MacLean et al. (1991) provide representations for design rationale as well as techniques for their capture. Design rationale techniques warrant further analysis by the software reuse community as one possible way forward.

Explanations based on criteria for component selection

Reusable code, design and specification components are often retrieved by a tool from a library or repository. Retrieval and selection criteria provide important knowledge which may be exploited during module explanation, as exemplified by explanation of legal cases in terms of their retrieval criteria in the intelligent legal assistant GREBE (Branting 1991). Explanation founded on matching criteria has received little attention in the reuse literature despite its intuitive appeal, as demonstrated by the following examples. In the income tax problem shown in Figure 3, well-defined selection criteria can be exploited to focus attention on the critical difference, for instance:

Module A is selected over Module B because Module A also matched on 'Previous-Year-Tax-Code'

Similarly explanation of the analogy between the air traffic control and flexible manufacturing specifications can be explained in terms of a generic domain class used to identify the analogical match. Indeed, experimental user studies of an intelligent reuse advisor has indicated that this domain class supports effective retrieval and effective explanation of analogical specification (Maiden and Sutcliffe, 1991). Simple explanation tactics such as description, justification and analogy were sufficient to ensure effective understanding of the analogical specification to inexperienced software engineers, see Maiden (1992). Two examples show that explanations founded on retrieval criteria provide a readily-available source of module descriptors, given of course that these retrieval descriptors are correct.

Comprehensive domain analyses

Explanation of reusable components based on their design rationale and retrieval criteria suggests the need for a partial analysis of the domains and applications behind reusable components. Software and knowledge reuse through domain analysis has been proposed elsewhere (Prieto-Diaz, 1990), primarily as a directly-reusable knowledge source. However, more comprehensive domain models representing semantically-rich application descriptions can be exploited to support reuse, for instance causal or structural relations between domain concepts can provide the basis for rich and causal explanations underlying software components. Domain modelling is likely to become a prominent reuse paradigm for the future, with particular emphasis on partial automation of software development, however, researchers should not lose sight of potential assistance for software comprehension provided by domain analysis.

Cognitive processes during reuse

Tool-based assistance for reuse can also be enhanced by knowing the skills and problems of software engineers during reuse. For instance, knowing the common errors made by software reusers can enhance tool support for reuse through inclusion of diagnostic facilities similar to those proposed in intelligent tutoring systems (Polson and Richardson, 1988). Similarly tool support is needed to encourage activities which software reusers do well. Two levels of tool-based support appear necessary for reuse. First, guidelines based on good software reuse practice should be encouraged among all reusers, so providing support at the behavioural and epistemic levels. Second, assistance at the individual level must recognise differences between individual reuser behaviour and errors. However, development of effective tool support underpinned by cognitive models of software reuse tasks remains a distant goal, requiring many comprehensive studies of human behaviour during software reuse.

10.3.2 Techniques to aid component comprehension and adaptation

This paper has identified six knowledge sources which may be exploited to assist software comprehension and adaptation. Providing explanations from these knowledge sources is one obvious technique for aiding component comprehension and adaptation, however additional techniques are also available and examined briefly in turn.

Controlled component transfer

Inhibiting the transfer and adaptation of reusable components until properly understood is one technique proposed in Maiden and Sutcliffe (in press). This controlled reuse is likely to discourage laziness manifest as copying and lead to more effective comprehension prior to customisation. For instance, strategies supporting specification reuse guided the reuser through major system processes, emphasising the importance of detailed reasoning about each area of the specification in turn. However, similar strategies supporting transfer of low-level design or code components may not be as effective, so different strategies may be necessary to guide reuse of varying software components.

Information hiding

Information hiding can focus attention on critical component features at the expense of other components. Furthermore, gradual exposure to the component combined with explanation dialogues based on knowledge discussed in the previous section can actively assist understanding. Information hiding should be integrated into controlled component transfer strategies outlined above, as was developed for the specification adaptation strategies described above. However, information hiding may have lesser or greater effects on the reuse of code or low-level design components, suggesting areas for further research.

Notetaking facilities

Notetaking facilities can assist comprehension by allowing software engineers to record critical facts about reusable components and similarities and differences between the new problem. Gick and Holyoak (1983, 1989) demonstrated the importance of graphical representations for understanding non-simple similarities between problem-solving situations, so support should permit both textually- and graphically-represented notes, as proposed by Haddley and Sommerville, (1990).

10.4 CONCLUSIONS

Human issues in software reuse have been ignored to a large extent in the literature. Human involvement is necessary but error-prone, for instance when defining new systems to be developed through reuse, understanding retrieved components and adapting them to fit new domains. Problems were demonstrated through six examples of reuse which emphasise difficulties often encountered by software reusers when reusing code, design and specification components. This paper suggests the need for both simple and complex tool-based support to alleviate some of these difficulties, based in part on the prototypical development of a support tool assisting software engineers to reuse specifications and designs (Maiden, 1992). Several of the simpler techniques discussed in this paper can be implemented using existing techniques and tools, although no detailed solutions and tool designs are proposed. Fischer and his colleagues in Colorado have also recognised the importance of human issues in software reuse, and have built several toolkits to investigate their hypotheses (Fischer et al., 1991a, 1991b). However, little other work has been reported, despite the potential pay-offs from software reuse. More human factors work is needed in the reuse community before truly effective reuse toolkits will be available.

REFERENCES

Byrne, E.J. 1991, Software Reverse Engineering: A Case Study, *Software – Practice and Experience,* **21** (12), pp. 1349-1364.

Chi, M.T.H., Bassok, M., Lewis, M.W. et al. 1989, Self-Explanations: How Students Study and Use Examples in Learning to Solve Problems. *Cognitive Science* **13**, pp. 145-182.

Fischer, G., Henninger, S. and Redmiles, D. 1991a, Cognitive Tools for Locating and Comprehending Software Objects for Reuse, Proceedings of 13th International Conference on Software Engineering at Austin, Texas, May.

Fischer, G., Henninger, S. & Redmiles, D. 1991b, Intertwining Query Construction and Relevance Evaluation, Proceedings of CHI'91, (eds.) Robertson, S.P., Olson, G.M. and Olson, J.S., ACM Press, pp. 55-62.

Furnas, G.W., Landauer, T.K., Gomez, L.M. and Dumais, S.T. 1987, The Vocabulary Problem in Human-System Communication, Communications of the ACM, **30** (11), pp. 964-971.

Gick, M.L. 1989, Two Functions of Diagrams in Problem Solving by Analogy, in *Knowledge Acquisition from Text and Pictures*, (eds.) Mandi, H. and Levin, J.R., Elsevier Science Publishers, B.V. (North-Holland), pp. 215-231.

Gick, M.L. and Holyoak, K.J. 1983, Schema Induction and Analogical Transfer, *Cognitive Psychology* 15, pp. 1-38.

Haddley, N. and Sommerville, I. 1990, Integrated Support for Systems Design, *Software Engineering Journal,* **5** (6), pp. 331-338.

Holt, R.W., Boehm-Davis, D.A., Schultz, A.C. 1987, Mental Representations of Programs for Student and Professional Programmers, in Empirical Studies of Programmers, Second Workshop, (eds.) Olsen, G. Sheppard, S. and Soloway E., Ablex, Norwood, NJ, pp. 33-46.

Jackson, M. 1983, *Systems Development*, Prentice-Hall International.

Langer, B.M. and Moher, T.G. 1989, Some Strategies of Reuse in an Object-Oriented Programming Environment, Proceedings of CHI'89, (eds.) Bice, K. and Lewis, C., ACM Press, pp. 69-73.

MacLean, A., Young, R.M., Bellotti, V.M.E. et al. 1991, Questions, Options and Criteria: Elements of Design Space Analysis, *Human-Computer Interaction*, **6** (3 and 4), Special Issue on Design Rationale (eds.) Carroll, J.M. and Moran, T.P.

Maiden, N.A.M. 1992, Analogical Specification Reuse during Requirements Analysis, Ph.D. Thesis under Submission, Department of Business Computing, City University, London.

Maiden, N.A.M. and Sutcliffe, A.G. (in preparation) The Abuse of Reuse: Why Software Reuse must be Taken into Care.

Maiden, N.A.M. and Sutcliffe, A.G. (in press) Analogically-based Reusability, to appear in *Behavioural Information and Technology*.

Maiden, N.A.M. and Sutcliffe, A.G. 1991, Analogical Matching for Specification Reuse, Proceedings of 6th Knowledge-Based Software Engineering Conference, Syracuse 22-25th September, pp. 101-112.

McCall, R. 1992, PHI: A Conceptual Foundation of Design Hypermedia, *Design Studies*, in press.

Mostow, J. 1989, Design by Derivational Analogy: Issues in the Automated Replay of Design Plans, *Artificial Intelligence,* 40, pp. 119-184.

Nanja, M. and Cook, R.C. 1987, An Analysis of the Online Debugging Process, in Empirical Studies of Programmers, Second Workshop, (eds.) Olsen, G. Sheppard, S. and Soloway, E., Ablex, Norwood, NJ, pp. 172-184.

Novick, L.R. and Holyoak, K.J. 1991, Mathematical Problem Solving by Analogy, *Journal of Experimental Psychology: Learning, Memory, and Cognition,* **17** (3), pp. 398-415.

Prieto-Diaz, R. 1991, Implementing Faceted Classification for Software Reuse, Communications of the ACM, **34** (5), pp. 88-97.

Prieto-Diaz, R. 1990, Domain Analysis: An Introduction, ACM SIGSOFT Software Engineering Notes, **15** (2), April, pp. 47-54.

Prieto-Diaz, R. and Freeman, P. 1987, Classifying Software for Reusability, IEEE Software, January, pp. 6-16.

Ross, B.H. 1989, Distinguishing Types of Superficial Similarities: Different Effects on the Access and Use of Earlier Problems, *Journal of Experimental Psychology: Learning, Memory and Cognition,* **15** (3), pp. 456-468.

Sutcliffe, A.G. and Maiden, N.A.M. 1990, How Specification Reuse can Support Requirements Analysis, Proceedings of Software Engineering 1990, (ed.) Hall, P., Brighton, UK, July 24-27, Cambridge University Press, pp. 489-509.

Thompson, R. and Huff, K.E. 1991, Supporting Understanding and Adaptation in Software Reuse, Proceedings 1st International Workshop on Software Reusability, Dortmund 3-5th July, pp. 45-50.

Watt, D.A., Wichmann, B.A. and Findlay, W. 1987, *Ada Language and Methodology*, Prentice-Hall International (UK) Ltd.

11 Process Modelling: A Critical Analysis

Anthony Finkelstein, Jeff Kramer and Matthew Hales
Imperial College, Department of Computing

11.1 INTRODUCTION

Software process modelling (aka process programming) has assumed consider-able importance in discussions of software engineering. In particular attention has been paid to the use of software process modelling in the construction of software development environments.

Despite the growing literature on this topic almost no independent critical analysis or evaluation has been available. This paper attempts to fill that gap. In particular we will be reflecting on experience with the Marvel environment from Columbia University. Marvel is the paradigm case of the software process mod-elling approach to building software development environments. In this paper we examine Marvel's strengths and limitations and look in detail at a small example of its use. We use this analysis as a basis for suggesting a research agenda for software process modelling.

11.2 WHAT IS SOFTWARE PROCESS MODELLING?

Essentially, software process modelling is the construction of an abstract description of the activities by which software is developed. In the area of soft-ware development environments the focus is on models that are enactable, that is executable, interpretable or amenable to automated reasoning.

A particular 'instance' of the software development process – the develop-ment of a particular piece of software – can be seen as the 'enaction' of a process model. That model can be used to control tool invocation and interwork-ing. A software development environment for a particular development is thus built up around (or generated from) an environment kernel which is essentially a vehicle for constructing and enacting such software process models. For a full discussion reference may be made to the extensive literature on this topic particularly to the Proceedings of the Software Process Workshop series (Potts, 1984; Wileden and Dowson, 1986; Dowson, 1987; Tully, 1989; Perry, 1990; Dowson, 1991).

The relation between software process modelling and reuse is complex. On the one hand models can be produced of software processes which incorporate reuse. An example of this will be discussed below. On the other hand software process models are themselves reused.

A software process model should encapsulate valuable organisational knowledge about the conduct of the development process. An example of this is the reuse of complex optimisation strategies (Wile, 1983). A further possibility is the construction of software process models in representation schemes that permit reuse, for example in object-oriented languages. These relations have been relatively little examined. All rest on an accurate assessment of the benefits and problems of software process modelling as a whole.

11.3 WHAT IS MARVEL?

Marvel is a software development environment kernel in the sense described above. An environment that has been produced with the Marvel kernel will be referred to as a Marvel environment. Marvel is the primary product of a research programme run by Gail Kaiser of Columbia University, New York. The goal of the programme is to 'develop a kernel for generating multi-user development environments that use knowledge about the software development process of large-scale projects'.

Marvel is based on experience with a multi-user programming environment called Smile (Kaiser and Feiler, 1987). In Smile the description of the programming process was 'hard-coded' into the environment. Marvel generalises this approach by providing support for the definition and enaction of software process models. Marvel environments run on top of the Unix family of operating systems.

There have been a number of versions of Marvel: version 0 was a 'proof of concept' implementation which simply replaced the Smile process description by a separate 'strategy' description; version 1 is the first large-scale implementation, based on the database manager developed for Smile; version 2 is a single-user implementation, independent of Smile, which includes a purpose built database manager; version 2.6, on which the example discussed below was based, includes some support for management of persistent data; Marvel 2.6 is supported on Sun-OS 4.0 with X-11 windows (also Ultrix 3.1 and AIX 2.2.1).

Marvel has been extensively described in the literature. Particular reference may be made to the key papers: Kaiser, Feiler and Popovich (1988) and Kaiser, Barghouti, Feiler and Schwanke (1988).

11.4 HOW DOES MARVEL WORK?

To construct a Marvel environment the developer must produce a data model and a process model. The data model describes the objects to be managed during the process of software development and their properties. The process model describes the activities carried out on those objects by the developers and tools involved in the specified development.

Marvel uses the data model to generate an 'objectbase' in which all artefacts (code, documentation, test cases and so on) created during a development are held. The objectbase also maintains history and status of the objects. The data model gives the types, or classes, of the objects involved, their attributes and the relationships between them. The objectbase is implemented straightforwardly as a Unix file structure. Each object instance has associated with it a unique directory, and directories are structured according to the relationships between the object instances.

The process model is given in the form of rules, each of which gives the preconditions which must be satisfied if the activity is to be carried out; the activity; and the postconditions, in terms of the effects of the activity on the objectbase (conditions and effects in AI planning system terminology).

The Marvel kernel provides a means of enacting the process model. It does so in an 'expert-system-like' manner by opportunistic processing. If the preconditions of an activity are satisfied that activity will be invoked, this may in turn result in the satisfaction of the preconditions of further activities and by forward chaining they will be invoked in turn. If a particular activity is chosen by a user and is not eligible for invocation the Marvel kernel will backward chain invoking those activities necessary for the selected activity to be performed.

Both data and process model are expressed as 'strategies' in the Marvel Strategy Language. Strategies can be imported into a main strategy. It is standard to define the data model in a single strategy but have multiple process model strategies for related tool sets.

11.5 INTEGRATING TOOLS WITH A MARVEL ENVIRONMENT

From the standpoint of an environment software development activities are performed by or through tools. In an open environment the appropriate tools for the development are added to the environment by the users. In the case of Marvel, and in fact with many other environments, the tools imported into the environment are raw Unix tools. To integrate a tool into a Marvel environment it is necessary to create an 'envelope' which will allow the Marvel kernel to invoke the tool with the correct parameters and, on exiting, the tool will define the results of its having been used.

11.6 THE TEST EXAMPLE

To test the viability of the overall software process modelling approach which Marvel exemplifies we attempted to build and use a Marvel environment for some example tools and build a test program in the environment we had constructed. The example, the Conic toolkit and a sample Conic application, is small but we feel illuminating.

11.6.1 The Conic toolkit

Conic is a distributed systems development toolkit developed by the Distributed Software Engineering Group at Imperial College. It supports the construction of distributed and concurrent programs by supporting the development of individual software components and the building of systems from these components. It provides tools for compiling and debugging individual components and supports the creation, linking and execution of these components (Magee, Kramer and Sloman, 1989).

Conic allows the definition of distributed systems by providing two languages: a programming language, for programming individual task modules (processes) with explicitly defined interfaces; and a configuration language, for the configuration of programs from groups of task modules. A variant of the configuration language is used to support dynamic creation, control and modification of these programs.

In Conic, context independent component types – group modules – can be defined within the configuration language and their interfaces given by typed entry- and exit-ports. Such component types may be built up from instances of other component types with the entry- and exit-ports linked to communicate by message passing. A system is a hierarchical structure of component instances in which the bottom level consists of task module instances written in the Conic programming language and executing concurrently. Nodes define the mapping between the component structure and the physical architecture. The programming language is Pascal extended to support message passing.

The Conic toolkit includes system development tools such as **cnc** (for invoking the compilers for both the configuration and programming language), and **ma** makefile generator (which uses the configuration language to establish dependencies). It also includes tools particularly appropriate for a distributed environment, such as **pb** playback allowing the previous execution of a node to be replayed from a trace. To manage an executing program tools such as create, link and remove are used. The **iman** tool gives a common interface to these tool fragments.

Considerable experience has been built up in the construction of Conic programs and a design method based on this experience developed (Kramer, Magee and Finkelstein, 1990).

11.6.2 A Conic application

The test application selected is extremely simple; it passes a character 'randomly' between a number of windows. The user determines the number of windows before running the application. When running the application the user enters in a starter window a character, a count and the number of the first window to receive the character and then the character is passed between the other windows until one window has displayed the character the determined number of times. The 'result' is then displayed in the starter window. It consists of two component types, starter and window, defined by group modules starter.grp and window.grp; starter.grp uses the task module controller.tas, and window.grp uses the task modules passer.tas and chooser.tas. All these modules use data type definitions held in a definition module window info.def, and chooser.tas also makes use of a library module to choose a 'random' window number. All components (.tas, .grp and .def) are separately compiled. An executive is included into a group module to enable it to be a distributable, runnable node.

This application was constructed in both the 'standard' Unix/Conic environment and reconstructed in the Marvel/Conic environment.

11.7 THE MARVEL/CONIC ENVIRONMENT

The simple environment constructed supports the editing, compiling and building of a system comprising a number of nodes and executable script files. It also supports a limited form of dynamic change.

To build the environment it was necessary to define the data model, process model and tool envelopes. These were produced by incremental and exploratory development.

The software objects to be stored in the objectbase are node, executive group, task and definition modules. These are organised into an SCENVIRONMENT which associates Conic application programs with executive modules available for use within a distributed node. Each application program is regarded as a SYSTEM.

An example object definition is shown below:

```
# A node is the unit of execution and distribution, and is the same as a
# group module except that it must contain an executive i.e. it must use
# an executive group module

NODE :: superclass ENTITY;
    name                  : string;
    availability_status   : (Available, NotAvailable) = NotAvailable;
    build_status          : (Built, NotBuilt) = NotBuilt;
    executive_used        : link EMODULE single;
    group_modules_used    : set_of GMODULE;
    task_modules_used     : set_of TMODULE;
    definitions_used      : set_of DMODULE;
    scripts               : set_of SCRIPT;        # Each script implicitly uses
end                                               # this node and the other
                                                  # scripts of this node
```

The process model developed was straightforward: modules can be created or modified by editing; imported definition modules must be available in the same directory as the importing module or in a directory in the user's search path; node modules can be compiled slightly differently from other modules, they may also be 'built' by compiling all the node components from scratch. Because activities can be performed on many objects we were able to overload the rules somewhat.

An example process model rule is shown below:

```
# The rule for compiling a task module

compile[?t:TMODULE]:

# Precondition

(forall DMODULE ?du suchthat (member [?t.definitions_used ?du]))
:
(and(?du.compile_status = Compiled)
     (?t.compile_status = ToBeCompiled))

# Activity

{ COMPILER compile_module ?t }

# Postcondition
(?t.compile_status = Compiled);
(?t.compile_status = ToBeEdited);
```

The tool envelopes were Unix shell scripts. A typical such envelope, the details of which are not of particular interest, is shown below to give a flavour:

```
# This script provides the compiler envelope for a module
# usage: compile_module module

echo $0 $1 ...
compile_prog=cnc
cd $1
files_used=`ls *_used/*.*/*.* 2>/dev/null`
if [ 'x$files_used' != 'x' ]
then
    # copy all files associated with modules used by this module
    # into module directory
    cp *_used/*.*/*.* . 2>/dev/null
fi
```

```
 module=`basename $1`
$compile_prog $module
if [ $? -eq 0 ]
then
    echo compile successful
    exit 0
else
    echo compile failed
    exit 1
fi
```

11.8 LESSONS LEARNED

In trying to give an account of the lessons learned from the construction of the simple environment discussed above we have attempted to filter out the lessons which are too Marvel specific, retaining those which we feel apply more generally to software process modelling as a whole.

11.8.1 Environment construction

The rigid distinction between the data model and process model is clearly impossible to maintain. In setting out the objectbase schema the attributes of objects form the potential pre- and postconditions of the process model. The two are closely intertwined.

Mapping Conic development to the objectbase was relatively straightforward, this seems to have been so because Conic systems have a clear structural foundation. The relation between the underlying representation scheme and the objectbase schema (and hence the process model) is very important.

Many of the tools we use in software development are highly generic. That is, we use them many times over during the development process but in different circumstances. The model case of this is the editor. We see relatively few activities (where activity translates to tool invocation) with many pre- and postconditions.

Because the way in which process models are developed is necessarily incremental and exploratory there is a strong requirement for support for analysis and debugging. The 'pure' process modelling aspects proved to be the least problematic aspect of constructing the environment.

The major problems we encountered were to do with building tool envelopes. We had to ensure our tools were 'well behaved' with respect to the data and process models. Clearly a major difficulty was the necessity of writing 'hacky' shell scripts (with all that this implies). Further we had to embed significant knowledge about the process and data model in the envelopes and vice-versa. There had to be a match between the postconditions specified by the process model and the exit status delivered by the envelope, hence further tying the envelopes and strategies together.

11.8.2 Environment use

In general, using the Marvel/Conic environment was straightforward. However a problem was encountered in what might be termed 'process over-automation'. Once an activity was invoked the user effectively lost control as the environment invoked further activities by chaining. It was, commonly, difficult to predict what activities might be invoked or their effects.

A specific difficulty arising from our use of the Marvel/Conic environment was data definition reuse. In contrast to C, Conic files cannot be compiled independently of external files which they reference. C files which are used by others need appear in only one position within the objectbase hierarchy whereas common Conic files are required during the compilation of any file in which they are referenced. This proves to be very difficult to handle with Marvel. It might be argued that this problem is not fundamental but is simply due to a bug or lack of functionality with Marvel. Certainly Marvel could be patched to eliminate the difficulty. More significantly however it again points to the close interlinking of objectbase structure, process model and structural aspects of the representation scheme. Similar problems were encountered with the handling of multiple instances of Conic components.

Setting the question of over-automation of the process aside, the use of the Marvel/Conic environment raises the issue of the granularity of the process model. In Marvel process control is at the level of tool invocation and interworking. Finer grain control could, in theory, be achieved, but only if the behaviour of the tool was very well understood (unlikely in the case of a 'foreign tool') and intermediate interaction with the objectbase could be defined. It remains an open question as to whether the level of granularity provided by a Unix or similar toolset is, except in strictly limited circumstances, the appropriate level for automation. To do so is to reduce a model of programming to an invocation of vi!

In the final analysis we were left with the feeling that we had gained relatively little for the effort of building the Marvel/Conic environment. With complex and high functionality tools such as cnc and ma the additional integration provide by the Marvel infra-structure seemed to add little except graphical display of environment status.

11.9 RESEARCH STRATEGIES

What are the implications of these observations for research strategy?

There has been a growing gap between research on the software development process and research on what might be called, for want of a better term, product representation schemes (specification languages and the like). Yet the concerns of the two strands of research have close parallels, for instance process modellers are concerned with coordinating the work of independent agents with differentiated roles and product specifiers with devising languages that provide for separation of concerns. This argues for research directed precisely at the point where process and product modelling are difficult to distinguish, in other words where the entities being manipulated are not anonymous 'objects' but are meaning-bearing elements of the underlying product language.

This argument also supports the need for fine-grain software development modelling, by which we mean the analysis and description of the detailed structure and organisation of development activities. In general this structure and organisation is ignored by modelling software development at the level of tool invocation and interworking. Indeed we suggest that many important gross features of software development such as verification, validation and cooperation arise from the complex interplay of fine-grain activities. These features of software development are not simply embedded in a matrix of routine 'house-keeping' tasks. Rather they are emergent properties that derive from the underlying fine-grain organisation. Fine-grain software development modelling may, as a by-product, build bridges between empirical (quantitative, experimental) studies of programmers behaviour and computational models.

The successful deployment of process modelling techniques is dependent upon what we have termed well-behaved tools. This means that the underlying tool interfacing and encapsulation techniques must be developed. If we envisage a distributed setting this poses important research challenges for tool implementors and for systems and language designers.

As is shown by our example above, the use of process modelling within software development environments is closely tied to the architecture and tool integration strategy of that environment. We are strongly antipathetic to environments based on global databases or centralised control through broadcast. A preference for loosely coupled and distributable architectures based on tool linking and specialised file storage stems both from 'software engineering instinct' and from experience with the difficulty of evolution and change in environments employing global integration schema. Further, we are strongly convinced that performance is of major importance in software development tools and environments. This again pushes us towards tool to tool integration (with direct transformations).

How exactly software process modelling fits within such an architecture is uncertain.

There is always a fine balance to be struck in providing support for software development activities between automation on the one hand and allowing direct intervention by the developer. How precisely that balance is made is highly dependent on the activities under consideration. Nevertheless we are inclined to the view that much existing process modelling research swings too far in favour of automation. An alternative is to view a process model as a vehicle for providing guidance to the developer so that at any time the user could ask 'What should I do next?' or perhaps 'How do I get out of this mess?'. In this setting the process model is more akin to computer-aided learning or tutoring than conventional software development support with all that this implies for research issues – what sort of model of the user needs to be preserved, how exactly should the guidance be presented. Further, it suggests that we look in general at interfaces and support for tool control and management.

Experience with software process modelling brings home forcibly the point that a 'generic' process modelling capability allows one to build 'stupid' processes. To use these techniques appropriately it is important that we have some handle on what a 'good' (economic, effective, uncertainty reducing) process would be.

Like any sort of 'programming', software process modelling is seductive. It is easy to get involved in the details and to set aside the question of whether or not the end product is of value. Most of the examples of software process modelling focus on configuration/version management, system building and test, yet these are areas in which we already have powerful and effective tools (RCS/SCCS/Make and so on), albeit that software process modelling may provide more genericity in their support. If software process modelling is to demonstrate its value it must do so in areas in which there is an identifiable gap in development support, this means generally up-stream in design and specification and with rigorous approaches to development.

A by-product of the seductiveness of enactable software process modelling is the dominance of 'environment' applications of software process modelling over its application in such areas as education (a traditional consumer of software process models) and process assessment both of which have received relatively little attention.

11.10 CONCLUSIONS

This paper has used Marvel and the Conic example as a framework for a critical analysis of software process modelling and has advanced an alternative (or at any rate complementary) research agenda in this area.

Many of the areas of research identified in this paper are also the subject of further work by the Marvel research team (Heineman, Kaiser, Barghouti and Ben-Shaul, 1991; Barghouti and Kaiser, 1991) and by others. A full review is beyond the scope of this paper but particular attention should be paid to the work of Feather, Fickas and Van Lamsweerde, for example Dardenne, Fickas and van Lamsweerde (1991).

A version of Marvel (3.0) is shortly to be released which provides support for multiple users. This version of Marvel, based on a client/server architecture, has a separate envelope language for providing interfaces to Marvel which is translated by a Marvel tool to sh, csh or ksh for execution. It also extends the objectbase implementation to support arbitrary relationships resulting in a directed graph rather than a tree. Other enhancements include tools for visual analysis of process models and built-in predicates to provide additional control on tool chaining.

The research agenda is the subject of further work by the authors (some preliminary results are given in Finkelstein, Kramer and Goedicke, 1990). We intend to extend our evaluation of software process modelling particularly to examine upstream development and an extended tool set.

ACKNOWLEDGEMENTS

We wish to acknowledge the contribution of Gail Kaiser and her team at Columbia University to the work reported above. Although in no way responsible for the opinions we have expressed, by licensing Marvel for use by Universities they have made it possible. Marvel licences are available for research purposes to educational and other non-profit institutions for a small processing fee (contact Prof. Gail E. Kaiser, Columbia University, Department of Computer Science, 500 West 120th Street, New York, NY 10027, USA, kaiser@cs. columbia.edu). All criticisms of the overall approach set aside, Marvel is an interesting product providing a clear and well-constructed demonstration of software process modelling techniques. It is a seminal contribution to the field and we would recommend it to any software engineering research group seeking to gain experience in this area.

Thanks also to our colleagues and students for their comments and technical assistance, in particular Jeff Magee, Manny Lehman and Kevin Twidle.

REFERENCES

Barghouti, N. and Kaiser, G. 1991, Scaling Up Rule-based Environments, Columbia University Technical Report, CUCS-047-90.

Dardenne, A., Fickas, S. and van Lamsweerde, A. 1991, Goal-directed Concept Acquisition in Requirements Elicitation, Proceedings of 6th International Workshop on Software Specification and Design, pp. 14-21, IEEE CS Press.

Dowson, M. (ed.) 1987, Iteration in the Software Process, Proceedings of the 3rd International Software Process Workshop, IEEE CS Press.

Dowson, M. 1991, Manufacturing Complex Systems, Proceedings of the 1st International Conference on the Software Process, IEEE CS Press.

Finkelstein, A., Kramer, J. and Goedicke, M. 1990, ViewPoint Oriented

Software Development, Proceedings of 3rd International Workshop Software Engineering and its Applications, Cigref EC2 1, pp. 337-351.

Heineman, G., Kaiser, G., Barghouti, N. and Ben-Shaul, I. 1991, Rule Chaining in Marvel, dynamic binding of parameters, 6th Annual Knowledge-based Software Engineering Conference; Rome Laboratory, New York, pp. 276-287.

Kaiser, G. and Feiler P. 1987, Intelligent Assistance Without Artificial Intelligence, 32nd IEEE Computer Society International Conference, pp. 236-241, IEEE CS Press.

Kaiser, G., Barghouti, N., Feiler, P. and Schwanke, R. 1988, Database Support for Knowledge-based Engineering Environments, IEEE Expert, 3 (3), pp. 18-32.

Kaiser, G., Feiler, P. and Popovich, S. 1988, Intelligent Assistance for Software Development and Maintenance, IEEE Software; 5 (3), pp. 40-49.

Kramer, J., Magee, J. and Finkelstein, A. 1990, A Constructive Approach to the Design of Distributed Systems, Proceedings of 10th International Conference on Distributed Computing Systems, May, pp. 580-587.

Magee, J., Kramer, J. and Sloman, M. 1989, Constructing Distributed Systems in Conic, IEEE Transactions on Software Engineering, SE-15 (6).

Perry, D. (ed.) 1990, Experience With Software Process Models, Proceedings of the 5th International Software Process Workshop, IEEE CS Press.

Potts, C. (ed.) 1984, Proceedings of the Software Process Workshop, IEEE CS Press.

Tully, C. (ed.) 1989, Representing and Enacting the Software Process, Proceedings of the 4th International Software Process Workshop, ACM SIGSOFT Software Engineering Notes, 14 (4).

Wile, D. 1983, Program Developments: formal explanations of implementations, CACM 26 (11), pp. 902-911.

Wileden, J. and Dowson, M. (eds.) 1986, Software Process and Software Environments, Proceedings of the 2nd International Software Process Workshop, ACM SIGSOFT Software Engineering Notes, 11 (4).

12 Writing Reusable Components in Ada

Mark Ratcliffe
Department of Computer Science, University College of Wales

12.1 INTRODUCTION

The development of the Ada programming language during the late 1970s and early 1980s, is a demonstration of the serious efforts undertaken by the software engineering community to encourage a more professional approach to software development. Ada must not be considered to be yet another programming language; it is all about software engineering and encompasses many of the modern techniques thought to be important in the development of software.

It should not be surprising to hear that Ada is an excellent vehicle to support software reuse; in fact it was designed that way. Packages and generics taken together provide the foundations necessary for the production of well-structured, reusable programs. Even before validated compilers had become available, many authors (Booch, 1983) had predicted the emergence of an Ada software components industry.

The model of reuse that Ada provides is essentially that of reusable packages, with each package providing a collection of closely related operations, often on a specific datatype. These packages are reused, without change, by encorporating them into new software modules. A number of examples of this type of reusable package library are described in Tafvelin, 1987.

Over the last ten years much work has been undertaken into how best to design Ada components to maximise their potential for reuse. This paper highlights some of the Ada Reusability Guidelines that were produced during the Alvey funded Eclipse project and described in detail in Gautier, 1990.

12.2 MODELLING REUSABLE COMPONENTS

One of the best examples of successful reuse in an engineering discipline is probably that of the electronic components industry Elliott, 1990. Here, the component represents a piece of hardware that has a set of clearly defined, physical interfaces that are available for connection to other components. The way in which these limited and precise interfaces are utilised is clearly stipulated through a set of well-defined protocols. This model of reuse is fundamental to the hardware industry and appears to work well. The similarities with the software industry are clear, and therefore we will pursue this model to determine whether it can be adapted to support the development and reuse of software components.

To follow the hardware analogy, and to maximise the potential reuse, we will add to our component a 'socket' through which it can provide services to other components. Similarly, for a component which wishes to use these specific facilities, we will define the requirements in the form of a 'plug'. To reflect the structure of these interfaces, plugs and sockets will be defined in many different shapes and sizes as defined by their specifications. When connecting the components together, the system will be policed in such a way that it will only be possible to insert a plug into a compatible socket.

In many cases, a single component may have both a socket and a plug. In order to provide services through the socket, such a component stipulates requirements which need to be satisfied through the accompanying plug before the socket can be used. An alternative method of dealing with such situations is to define a sub-component that can provide the required resources internally; such a design decision can seriously reduce the reusability of the components involved. A component which offers the facility of a binary tree, for example, may provide a socket to export procedures such as INSERT and SEARCH and, in order to maximise reuse, is likely to require a plug to ascertain the type of objects that are to be imported into the tree. If the type of this object is satisfied by an internal sub-component providing integers, for example, then the effective reuse of that component is reduced.

It should be noted that no stipulation is being made here as to the direction of flow of information across plugs and sockets.

12.3 APPLYING THE MODEL TO ADA

Within the Ada programming language, our analogy to the hardware component is realised in the form of the package. This construct allows a data structure and the operations which can be performed upon it to be combined into a single unit; the actual implementation of the data structure and the associated operations are kept separate and are maintained within a package body. The package body is not visible to the user of a package.

By providing a more powerful and more flexible form of abstraction than can be achieved by functions and procedures alone, the package is able to provide a framework for overcoming the problems of complexity inherent in large systems. Furthermore, by separating out the implementation details and keeping them hidden from the user, the package construct maximises the level of reusability provided by a component.

The user's view of a component is restricted solely to that made visible in the package specification. This specification represents the 'socket' of the component and includes the definition of all operations and types that the component exports.

In addition to improving the reusability of a component, the separation of implementation and specification has the added advantage of allowing the implementation to be changed without affecting the external interface of the component. Unless there are specific requirements to the contrary, such as performance and resource usage, users of a component should be unaware of any changes that are made to an implementation.

```
package STACKS is
    type STACK is limited private;
    procedure PUSH    (S : in out STACK; X : in      INTEGER);
    procedure POP     (S : in out STACK; X :    out INTEGER);
    function    DEPT        (S : STACK) return INTEGER;
    function    IS FULL     (S : STACK) return BOOLEAN;
    function    IS EMPTY  (S : STACK) return BOOLEAN;
private
    type STACK is ...;
end STACKS;
```

Figure 12.1 The 'stacks' package

Figure 12.1 shows an Ada package that provides a basic stack routine; unfortunately though the interface is clearly defined, the component is not, as yet, very reusable, unless of course there is a specific requirement for a component that supports a stack of integers.

We have seen that a component becomes more reusable if the facilities it offers are clearly identifiable. This concept can be extended by also applying it to the requirements, or 'plug', of a component. Almost all components have requirements of some form or other that are to be satisfied by other components. A compiler, for example, needs a lexical analyser, which may itself use a tree package, and so on. By capturing these requirements and representing them as identifiable interfaces of the component, then clearly a component's potential for reuse can be increased. For example, if it becomes necessary to improve the performance of a compiler, assuming that its component parts are well defined, some of the components that it uses could be substituted; an inefficient optimiser could be replaced with a better one.

Requirements such as those just described may be considered to be the concern of the implementation and as a consequence be supported by means of the 'with' context clause. This is too restrictive as it binds an internal plug of that component to a particular socket of the other 'withed' component.

In Ada, we are able to support component plugs by the use of generics. This construct improves the flexibility and adaptability of the package by providing a template which may be parameterised. The generic component can then be instantiated by a user to create a customised subprogram or package. This idea, of creating a general interface to a component and then allowing the users to customise it to suit their own application, is vital to reuse.

```
generic  — define generic parameters for STACK
   type ITEM is limited private;
package STACKS is
   type STACK is limited private;
   procedure PUSH  (S : in out STACK; X : in    ITEM);
   procedure POP    (S : in out STACK; X :    out ITEM);
   ...
end STACKS;
```

Figure 12.2 The generic 'stack' package

Figure 12.2 shows how the stack package can be constructed to support a generic parameter. In this example, ITEM represents a generic type which can be instantiated with a type of the user's choice to create a specific Stack package. If, at instantiation time, ITEM is replaced with INTEGER, then an equivalent package to that shown in Figure 12.1 is created.

The generic package provides the software engineer with the programming equivalent of the hardware component. Unfortunately, the major problem with this approach is that Ada does not allow a package to be passed as a single generic formal parameter; all that is permitted is a flat list of declarations, in effect a single large plug. This problem can be alleviated to a certain extent by the use of blank lines, comments and indentation in the layout of the generic parameters in order to group related items into individual plugs.

Having established a mechanism for our 'plugs' and 'sockets', we will now consider design techniques which will help maximise the reuse of our components.

12.4 DESIGNING THE COMPONENTS

To be truly reusable, our component needs to be designed in a way which encompasses both the immediate and long-term needs. Interfaces should be designed that are suitable for a general model rather than for the immediate need.

We have seen that Ada supports abstraction through the use of the package, however, the reuse guidelines suggest that this is not enough; packages, in fact, support two types of abstraction based upon the methods used to maintain state information that is saved from one operation to another. This information may be held internally in variables declared in the private part or body of the package, in which case the package is referred to as an Abstract State Machine (ASM). Alternatively, a package may export a type declaration so that the user can declare objects to hold the information. This form of package is referred to as the Abstract Data Type (ADT) (Guttag, 1980) and comes close to the idea of Object Oriented Programming.

The guidelines give the example shown in Figures 12.3 and 12.4 to explain the merits of the two different approaches.

```
generic
    type ITEM is (<>);
package DISCRETE_SET_ADT is
    type SET is private;
    procedure INSERT        (I : in ITEM; INTO : in out SET);
    function IS_MEMBER  (I : ITEM; OF SET : SET)
                                return BOOLEAN;
end DISCRETE_SET_ADT;
```

Figure 12.3 SET package as an ADT

```
generic
    type ITEM is (<>);
package DISCRETE_SET_ASM is
    procedure INSERT        (I : in ITEM);
    function IS_MEMBER (I : ITEM) return BOOLEAN;
end DISCRETE_SET_ASM;
```

Figure 12.4 SET package as an ASM

The main difference between the two methods becomes apparent when one tries to build a new operation for each of the two packages; for example, to implement the UNION operator in the ADT, the code shown in Figure 12.5 is required.

```
package CHAR_SETS is
   new DISCRETE_SET_ADT (ITEM => CHARACTER);
use CHAR_SETS;          -- for document convenience only

function UNION (S,T : SET) return SET is
   RESULT : SET;
begin
   for CHAR in CHARACTER loop
      if IS_MEMBER   (CHAR, OF_SET => S) or else
         IS_MEMBER   (CHAR, OF_SET => T) then
               INSERT    (CHAR, INTO => RESULT);
      end if;
   end loop;
   return RESULT;
end UNION;
```

Figure 12.5 UNION for the ADT package

This works well, but when we try to repeat a similar operation using the ASM package it becomes necessary to represent the operation as a generic procedure, as is shown in Figure 12.6. This is because Ada packages cannot be passed as parameters to functions, procedures, or generics. This could be instantiated and called as in Figure 12,7.

```
generic -- procedure UNITE
   type ITEM is (<>);
   with function   IS_MEMBER_A   (I : ITEM) return BOOLEAN;
   with function   IS_MEMBER_B   (I : ITEM) return BOOLEAN;
   with procedure  INSERT_R       (I : in       ITEM);
procedure UNITE;

-- generic body
procedure UNITE is
begin
   for I in ITEM loop
      if IS_MEMBER_A (I) or else IS_MEMBER_B (I) then
         INSERT_R (I);
      end if;
   end loop;
end UNITE;
```

Figure 12.6 UNION for the ASM package

```
declare
    package A is new DISCRETE_SET_ASM ( ITEM => CHARACTER);
    package B is new DISCRETE_SET_ASM ( ITEM => CHARACTER);
    package R is new DISCRETE_SET_ASM ( ITEM => CHARACTER);
    procedure R_BECOMES_A_UNION_B is
        new UNITE (ITEM => CHARACTER,
                   A.IS_MEMBER,
                   B.IS_MEMBER,
                   R.INSERT);
begin
    R_BECOMES_A_UNION_B;
    ...
```

Figure 12.7 Instantiation of ASM UNION procedure

Another major problem of using packages which maintain internal state occurs when an attempt is made to use such a package in a multi-tasking environment. This is likely to produce unpredictable results. On the other hand, a package implementing an abstract data type will not cause such problems as the responsibility for controlling access to the shared state lies solely with the reuser.

As described in the reusability guidelines, we should avoid specifying a package in such a way that the implementations of that package have to maintain state.

12.4.1 Sufficient, complete and primitive qualities

In general, a component which has a single, simple interface is more reusable than one with a greater number of more complex interfaces. However, the interfaces must be adequate to provide the required generality of the function. This requirement on interfaces is often described as being 'sufficient, complete and primitive'.

A component is described as being 'sufficient' if it includes all those operations considered to be capable of handling the object. A STACK component, for example, is not usable unless it has sufficient operations available to both insert and retrieve data – it must have both PUSH and POP available.

A component is described as being 'complete' if it includes all of those operations which characterise the object. Whereas sufficiency implies the smallest collection of meaningful operations, a complete component is one in which all aspects of the underlying abstraction are covered. There is a danger that this completeness property can cause the component to grow very large and for this reason components are also required to contain only 'primitive' operations.

Primitive operations are those which require knowledge of the underlying implementation in order to be implemented efficiently. Deciding on whether an operation is primitive is not straightforward. At first it may appear that a function to COPY a stack is not a primitive as the operation may be achieved by utilising a temporary stack and using a sequence of PUSH and POP operations. Such a process would however temporarily alter the state of stack being copied; for this reason and for the sake of efficiency it is usually considered to be primitive.

In summary, a component becomes usable when it is sufficient, becomes extensible when the abstraction is primitive and becomes applicable to a wider range of applications when it is complete.

12.4.2 Cohesive qualities

To maximise its reusability, a component should also exhibit low coupling and high cohesion. Cohesion refers to the degree to which the operations provided by a component are closely related; thus a component which combines a function for determining the size of a linked list together with a procedure for printing out the date, demonstrates low cohesion. A component which simply provides the PUSH and POP operations on a stack demonstrates high cohesion.

The extent to which a component depends on other components is referred to as coupling. A component which imports other components becomes less reusable as more of the environment has to be brought in along with the component itself.

Ideally a component should not exhibit coupling and should be completely independent of all other components. Few components are likely to offer such properties and those that do are unlikely to be of much use. As already described, the more satisfactory solution is to have all dependencies established at instantiation time by the use of generic formal parameters.

12.5 DEALING WITH EXCEPTIONS

Designing our components in a way which maximises their reuse seems to fit well into our original hardware model. Unfortunately, Ada does provide a number of obstacles to the straightforward use of our plugs and sockets; the problem of passing packages has already been discussed. Another problem lies with the exception construct which, though often provided as a socket, cannot be plugged. Careful design considerations can be used to overcome this weakness.

Exceptions are not only provided for dealing with exceptional events, they are also commonly used to signal erroneous or unexpected conditions. Consider, for example, the reading of a file; if that file is not already open for reading, such an operation will raise a STATUS exception. In this case the exception could have been, and perhaps should have been, tested for. There are certain conditions, such as a hardware DEVICE upon writing to a file, which cannot be tested for.

In the case of exceptions, the reusability guidelines do not dictate a particular programming style, but rather recommend that for each condition required for the successful operation of a subprogram, an exception should be defined to be raised when that condition does not hold. In addition, wherever possible, functions should be provided to enable a user to test for the conditions under which the exceptions exported by that package are raised.

The most important aid to the support of exceptions is that of documentation. Consider the example shown in Figure 12.8.

```
with IO_EXCEPTIONS;
use IO_EXCEPTIONS;  -- for document convenience only

package TEXT_IO is
   ...
   procedure OPEN   (FILE : in out FILE_TYPE;
                     MODE : in    FILE_MODE;
                     NAME : in    STRING;
                     FORM : in    STRING := "");
   — raises  STATUS_ERROR when IS_OPEN (FILE)
   — raises  NAME_ERROR   when "NAME does not identify file"
   — raises  USE_ERROR    when "MODE and FORM are not
                                supported by file NAME"

   procedure   CLOSE (FILE : in out FILE_TYPE);
   -- raises    STATUS_ERROR when not IS_OPEN (FILE)

   function    IS_OPEN (FILE : in FILE_TYPE) return BOOLEAN;
   ...
end TEXT_IO;
```

Figure 12.8 Documenting exceptions

The specification shows that a particular operation can raise specified exceptions. It is not clear, however, whether these are the only exceptions which can be raised by an operation. In Ada, the standard way of dealing with exceptions is to propagate up those exceptions not handled by an operation. By default, we will therefore assume that an operation can propagate 'others' unless otherwise documented.

It is not necessary to become over defensive when designing components; a component need only handle those exceptions which are raised in directly used components. Providing that the used components are also classed as reusable, they in turn should not knowingly propagate anonymous or implementation-specific exceptions.

12.6 CONCLUSION

The Ada programming language does much to support software reuse; it is the author's belief that the language is quite adequate to allow reuse to be exploited profitably. However, for reuse to become an integral part of the software development process requires the development of a 'reuse culture', in which the search for existing components instinctively precedes any attempt to design new ones. The main obstacle to reuse now appears to be organisational (Tracz, 1988) but until there is a strong commitment to achieving reuse, it is unlikely to become common place.

Finally, we should consider whether it is feasible for a software components industry to develop so that it is capable of supplying components to suppliers of complete software systems. This idea seems to have been first conceived by McIlroy (McIlroy, 1969); a fanciful version is to be found at the end of Barnes' book on Ada (Barnes, 1982).

There is now at least one company in the USA which is marketing a set of reusable components in this way. The components commonly used are data structures. They are provided in source form and are made available on the basis of a site licence which allows them to be copied and used freely on that site; furthermore, the components can be incorporated into complete systems and sold on without any royalty being due to the component supplier. The components are guaranteed so that any defect reported will be rectified free of charge; there is therefore no maintenance charge. It remains to be seen just how successful this scheme will be.

BIBLIOGRAPHY

Barnes, J.G.P. 1982, Programming in Ada, Addison-Wesley, Wokingham, UK.

Booch, G. (ed.) 1983, Software Engineering with Ada, Benjamin Cummins, California, USA.

Elliott, A., Gautier, R.J. and Welch, P.H. 1990, Component Engineering in Ada – Some Problems and Some Advice, in Gautier, R.J. and Wallis, P.J.L. (eds.) *Software Reuse with Ada*, Peter Peregrinus, UK.

Gautier, R.J. and Wallis, P.J.L. 1990, *Software Reuse with Ada*, Peter Peregrinus, UK.

Guttag, J. 1980, Abstract Data Types and the Development of Data Structures In Programming Language Design, Los Alamos, California, USA, IEEE Computer Society Press.

McIlroy, M.D. 1969, Mass Produced Software Components, in Naur, P.,

Randell, B. and Buxton, J.N. (eds.) Proceedings of NATO Conference on Software Engineering, New York, Petrocelli/Charter.

Tafvelin, A. 1987, Ada Components: Libraries and Tools, In Ada-Europe International Conference, Cambridge University Press.

Tracz, W. 1988, Tutorial: Software Reuse: Emerging Technology, IEEE Computer Society Press, USA.

13 AD/Advantage – A Practical Software Reuse Solution

Gordon Woodcock
Cincom Systems UK Limited

The dream of IT is to be able to automate both the development and maintenance of IT systems. We can draw an analogy with the idea of a machine, where ideas for new systems, or changes to existing systems, float into a hopper, we crank the handle and finished business applications come out of the other end. Unfortunately, reality often falls short of dreams.

The reality is that lots of work is involved. Sophisticated commercial and industrial applications are being developed by many companies, and the complexity involved can be quite daunting. It is also still commonplace for this work to be primarily manual. In a lot of cases the resulting application is not fully in line with corporate goals – in other words is not exactly what is required, – was not on time, and falls short of the desired quality. Although the IT industry has the dream, it also recognises the problems.

The key problem is that there is a great divide separating the people with the design information from the finished software product. The great divide is filled with the dangers of complexity, fast changing technical environments, and the fast changing world of business. Companies are driven hard by the need to support corporate goals and achieve competitive advantage. We've learnt the hard way not to underestimate these problems.

Dr F.P. Brooks is the father of the System 360 from IBM, he states that 'Software entities are more complex for their size than perhaps any other human construct'. This highlights the difficulties we have been discussing. It is tempting, but misleading, to believe that CASE tools provide a complete solution to these problems.

Generating programs from CASE design information is a first step to bridging the great divide and realising the dream. However the industry realises that before the design is frozen into the finished code there are several further stages of development still needed. A process of continuous refinement is necessary – for reasons of modelling, feasibility studies, prototyping, achieving better quality and high performance, or simply improving the application ergonomics. Other industries have learned how to address this issue.

The aircraft industry faces identical problems. They do not take a blueprint design and go straight into building the real thing. They build models, prototypes, test planes, simulate the environment with wind tunnels, and so on. Why do they do this? Because it is too expensive to make mistakes in the final product.

In September 1989 IBM launched AD/Cycle. Not a product, but a framework for a set of products – whether IBM or third party products, IBM recognise the modern thinking for software development and maintenance, and I applaud them in this. The benefits will be appreciated by users, by IBM and by third party vendors. The productivity gains that are possible with new technologies is also something that manufacturing industry is familiar with. For example, hand tools provide capabilities of mechanical leverage for us. Electrical/hydraulic/steam power when applied to those tools provides a quantum leap in productivity. This is repeated yet again by the application of computer capabilities, to jump to the latest level of technology-automated manufacturing, or robots.

So the manufacturing industry has two areas of technical achievement; CAD-computer aided design, CAM-computer aided manufacturing. The next step is to integrate these islands of automation – to feed the designs achieved in CAD into the CAM process, to achieve a further quantum leap in productivity. We can relate this to the IT industry and realise that 'CASE' as we know it is really CASD – computer aided software design. So how can we achieve those same leaps in productivity in software production?

AD/Advantage automates the process of software development by generating the system components from the design tool. The finished system is then produced through a process of gradual refinement. So we have a close coupling of CAD and CAM techniques for the software world.

As with the manufacturing world, from hand tools to robots, the evolution of technology is a series of disjoint progressions. As software technology has advanced from manual cobol to generators and 4GL systems, AD/Advantage is the next progression. Conservative figures from CINCOM development projects have shown approximately 4:1 improvement over 4GL technology. This is equivalent to a 16:1 improvement over the manual use of COBOL. How does AD/Advantage achieve this?

The first process is one of automatic code generation. The design data is used to generate the framework or structure of the system automatically. The detail design data is used to generate the lower level components of the application, or building blocks. This process of forward engineering from a design source is the first stage. The next stage is to assemble these components. The generated components or building blocks are maintained within a reusable code management system.

The system of reuse has some far reaching benefits which permeate the entire process design. No longer is every application built from scratch. Instead you will reuse a component many times in a single application, and will reuse components from previously built applications. So a finished application could include 15,000 lines of reused code for every 10,000 lines of newly generated code. A significant gain. Maintenance workload is a larger problem area than new development for most companies. As maintenance of existing systems grows, it can choke the ability to develop new systems. The system of reuse is designed to resolve these problems. A change in one component or building block will ripple the new version throughout the system. The control and productivity gains are again dramatic. Also all changes are reversed back to the design source; these features all have one other benefit in common – Quality.

High quality of the finished product is obtained in two ways. Quality is created initially through the generation of the application framework and components directly from the design tool. This quality is enhanced through the refining and testing of the composed application. This quality is maintained and propagated through the system of reusable components, ensuring that all applications benefit, and quality will not degrade over time.

Using these techniques it is no longer necessary to generate whole programs or systems in one go. It is not an all or nothing step, instead the final application is derived through a process of gradual refinements. Several techniques can be used, prototyping, modelling, hands on testing etc. These techniques mean that quality is built in, and the final applications can equally meet the goals of the corporation.

Corporate applications also have to look to the future technologies. With this objective in mind, AD/Advantage is built with open systems in mind. Only the business application is strategic to the customer. In all other areas such as hardware and software, the choice should be open. AD/Advantage is designed with common standards in mind – AD/Cycle, Repository, Cobol, platform independence to facilitate this. Also AD/Advantage prevents any proprietary lock-in to itself, by provision of reverse engineering, repository access and cobol generation. This gives the customer choice and flexibility, without future lock-in.

Contributors

Anthony Finkelstein, Jeff Kramer and
Matthew Hales
Imperial College
Dept of Computing
180 Queens Gate
London
SW7 2BZ

Malcolm Fowles
James Martin and Co
11 Windsor St
Chertsey
Surrey
KT16 8AY

Pat Hall
Open University
Dept of Computer Science
Walton Hall
Milton Keynes

Ivan Kruzela
Telia Research
Box 85
20120 Malmoa
Sweden

Neil Maiden
Dept of Business Computing
City University
Northampton Square
London
EC1V 0HB

Patrick McParland
Dept of Computer Science
The Queen's University of Belfast
Belfast
BT7 1NN

David Mole
Southbank Polytechnic
CSSE Room 401
Borough Rd
London
SE1 0AA

Mark Ratcliffe
University College Penglais
Aberyswyth
Dyfed
Wales
SY23 3BZ

Ian Reekie
Instrumatic UK
First Avenue
Marlow
Bucks
SL7 1YA

Alistair Sutcliffe
Dept of Business Computing
City University
Northampton Square
London
ECIV 0HB

Paul Walton
MacDonald Dettwiler
Branksome Chambers
Branksome Rd
Fleet
Hampshire
GU13 8JS

Martin Ward
Centre for Software Maintenance Ltd
Unit 1P
Mountjoy Research Centre
Stockton Road
Durham
DH1 3SW

Gordon Woodcock
Cincom Systems (UK) Ltd
Crown House
Manchester Rd
Wilmslow
Cheshire
SK9 1BH